44.95

Materials Selection for the
Chemical Process Industries

Other McGraw-Hill Chemical Engineering Books of Interest

AUSTIN ET AL. • *Shreve's Chemical Process Industries*

COOK, DUMONT • *Process Drying Practice*

CHOPEY • *Handbook of Chemical Engineering Calculations*

DEAN • *Lange's Handbook of Chemistry*

DILLON • *Materials Selection for the Chemical Process Industries*

FREEMAN • *Hazardous Waste Minimization*

FREEMAN • *Standard Handbook of Hazardous Waste Treatment and Disposal*

GRANT, GRANT • *Grant & Hackh's Chemical Dictionary*

KISTER • *Distillation Operation*

MCGEE • *Molecular Engineering*

MILLER • *Flow Measurement Engineering Handbook*

PERRY, GREEN • *Perry's Chemical Engineers' Handbook*

REID ET AL. • *Properties of Gases and Liquids*

REIST • *Introduction to Aerosol Science*

RYANS, ROPER • *Process Vacuum Systems and Design Operation*

SANDLER, LUCKIEWICZ • *Practical Process Engineering*

SATTERFIELD • *Heterogeneous Catalysis in Practice*

SCHWEITZER • *Handbook of Separation Techniques for Chemical Engineers*

SHINSKEY • *Process Control Systems*

SHUGAR, DEAN • *The Chemist's Ready Reference Handbook*

SHUGAR, BALLINGER • *The Chemical Technician's Ready Reference Handbook*

SMITH, VAN LAAN • *Piping and Pipe Support Systems*

STOCK • *AI in Process Control*

TATTERSON • *Fluid Mixing and Gas Dispersion in Agitated Tanks*

YOKELL • *A Working Guide to Shell-and-Tube Heat Exchangers*

Contents

Preface

When the author first became involved with corrosion, materials selection, and failure analysis in the chemical process industries, there were no formal educational courses in such subjects at the undergraduate level. There was no basic corrosion course, such as is provided by NACE (the National Association of Corrosion Engineers) today, nor the ASM International course on corrosion.

Fortunately, there appeared in 1950 a most useful book. This was *Materials of Construction for Chemical Process Industries* by James A. Lee, Southwestern editor of McGraw-Hill Publications, formerly managing editor of *Chemical Engineering* and its predecessor *Chemical and Metallurgical Engineering*. This book, a volume in the McGraw-Hill Series in Chemical Engineering, addressed itself to materials of construction for the production, handling, and packaging of more than 300 chemicals or combination of chemicals. Each chemical or combination of chemicals was handled as a separate chapter (although some were very brief) and arranged in alphabetical order.

Lee's book was invaluable to the starting corrosion engineer and the engineer's clientele at the time. It is, however, long out of print. Furthermore, the processes are archaic in many instances, as are the materials of construction.

The present author has tried to provide a new reference work in the spirit of Lee's effort, albeit in a slightly different format. Discussing the principles of materials selection, heavy chemicals manufacture, inorganic preparations, selected organic syntheses, and specific problems associated with the chemical process industries, I have prepared a reference work which I hope will be useful to design, operating, and maintenance personnel as well as the practicing corrosion or materials engineer. I have relied heavily on the Kirk-Othmer *Encyclopedia of Chemical Technology*, 3d ed. (John Wiley & Sons, New York, 1978) for process information, tempered with input from current industrial contacts.

The author welcomes constructive criticism and suggested material for inclusion in future editions. I sincerely hope that the corrosion/materials engineering community will not let another three or four decades go by without an appropriate update by experts.

C. P. Dillon

Introduction

This book is intended both for the professional corrosion/materials engineer and for the design, operations, or maintenance engineer concerned with process industry plant (e.g., chemical, petrochemical, or fertilizer). It will provide suggestions for materials selection for the processes described, warnings about pitfalls associated with specific chemicals or mixtures of chemicals, and measures suggested to cope with certain specific problems associated with process industry corrosion.

The author has organized the text into five sections. These are:

Part 1—Materials Selection

Part 2—Heavy Chemicals

Part 3—Inorganic Processes

Part 4—Organic Syntheses

Part 5—Special Topics

Within Parts 2, 3, and 4, simplified flow diagrams are provided for one or more of the most common processes for the chemical of interest.

The chapters in these sections are arranged alphabetically for ease of reference, on the premise that the engineer is concerned with materials for a specific chemical or process rather than what to do with a specific material of construction. Nevertheless, a brief index has been provided to facilitate searching for materials or phenomena.

Each chapter is arranged to present first the relevant process and suggested materials of construction, then information on handling the chemical for shipment and storage. The pitfalls associated with production or utilization are discussed (e.g., effects of specific contaminants, phenomena typical of the exposure). Last, there are comments on the relevant materials of construction themselves.

When discussing specific alloys, the designation used is that of the Unified Numbering System (UNS), fifth edition [American Society for Testing and Materials (ASTM) DS 56 D]. For nonmetallic materials, the accepted designation is used (e.g., PTFE for polytetrafluorethylene polymer).

The author has attempted to indicate appropriate materials or groups of materials which might be employed, as indicated by the published literature or by private communication with knowledgeable people in the industry. Information from previously published articles is acknowledged but is edited to an updated form, to include new developments or improved materials.

The information is provided as a point of departure for final materials selection, which must always be predicated on expected life, cost/benefit ratios, and availability, as well as the economic parameters peculiar to the industry or company concerned. It is recognized that materials costs and availability will vary from one country to another, that tax structures and accounting procedures will vary, and that policies regarding the use of indigenous materials (or high tariff rates on imported materials) may militate against the effective use of a suggested material.

Last, it should be recognized (as described further in Chap. 3 particularly) that there is no way in which this book could anticipate all possible variations or fluctuations in a given process that might arise from contaminants in the feed streams, from recycling process streams, etc.

This work is therefore intended as a useful guide to proper materials selection in the chemical process industries but cannot be used as a "cookbook," despite the best efforts of the author, reviewers, and contributors.

Materials Selection

2

Principles of
Materials Selection

Basically, the procedure for materials selection entails five steps. These are as follows:[1]

1. Defining the conditions of exposure and the consequent requirements for each item of equipment.
2. Establishing the strategy for evaluating candidate materials.
3. Identifying the candidate materials.
4. Evaluating the materials of construction.
5. Selecting the final materials, with the necessary documentation and communications.

As described further below, the recommended sequence for following these steps lies in the definition of technology, definition of facilities, and attendant procedures related to construction, start-up, and operations and maintenance.[2]

The responsible engineer will be aware of the major factors affecting materials selection: physical and mechanical properties, regulatory codes, fabricability, corrosion characteristics and amenability to corrosion control, reliability and safety, and economics. The major effort in materials selection is made in the design stages. Nevertheless, there will be a constant need for attention in the operation and maintenance of plant, including the reassessment of performance of existing materials and the possibility of upgrading equipment performance through the use of improved materials or corrosion control techniques.

The optimum materials are those which have the lowest cost/life ratio when fully defined as the annual cost in consideration of tax rates, the value of money, and appropriate accounting procedures.

Definition of Technology

The definition of technology (DOT) details the process chemistry and conditions. It relates primarily to new reactions developed in research and development. It is mentioned here because an apparently established process may have new developments in regard to catalysts, contaminants (e.g., from recycling streams, because of environmental considerations), or conditions of temperature, pressure, or concentration—developments which may profoundly affect the materials selection.

A preliminary materials selection (subject to review and change as described further below) can be based on

1. The major and minor constituents of the process streams, including trace contaminants, pH (or other measure of acidity/alkalinity), and degree of aeration
2. Process conditions
   a. Temperature (normal, maximum/minimum, possible excursions, shutdown effects, rates of heating or cooling)
   b. Pressure (normal, maximum/minimum, excursions, sudden drops leading to cavitation or turbulence)
3. Velocity effects (laminar vs. turbulent flow, impingement, erosion, abrasion, or cavitation)
4. Contaminants (e.g., chlorides from condenser leaks, alkaline species carried over in steam, specific metal ions from catalysts or corrosion)

Definition of Facilities

The definition of facilities (DOF), sometimes known simply as the *process design package,* details the necessary numbers, kinds, and sizes of major items of equipment and their physical layout at the plant site. Note that the preliminary materials selection may require revision in the light of specific design considerations or plant site conditions (e.g., cooling water quality).

The DOF should include process and control diagrams (P&CDs), plant layout diagrams, major equipment lists, line lists, and pertinent engineering criteria, standards, and specifications.

Construction, Start-up, and Operation

The DOF documents are essential to the construction, inspection, and operation and maintenance of the plant.

A simplified flow diagram, along the lines of those in this book, is helpful for operation and maintenance personnel concerned with maintaining and refurbishing the equipment. The flow diagram should show the original materials of construction as chosen usually by the materials engineer (with applicable ASTM or other specifications) and must be revised periodically, as required.

There must also be adequate documentation of ancillary requirements related to process waters, wastewaters, and atmospheres (e.g., inert gas requirements) which might adversely affect the operation and/or product quality.[3]

References

1. R. B. Puyear, "Materials Selection Criteria for Shell-and-Tube Heat Exchangers for Use in Process Industries," *Shell and Tube Heat Exchangers,* American Society for Metals (ASM), Metals Park, Ohio, 1980.
2. C. P. Dillon, *Corrosion Control in the Process Industries,* McGraw-Hill, New York, 1986.
3. C. P. Dillon, "Eight Items Affect Materials Choices," *Oil and Gas Journal,* Oct. 10, 1977.

Chapter

3

Oxidizing and Reducing Environments

In an attempt to simplify the corrosion picture as to probable behavior of engineering materials in different environments, the corrosion community may have obscured some basic facts. We refer to the Y diagram (Fig. 3.1) and similar representations that relate the corrosion resistance of metals and alloys to "oxidizing" and "reducing" environments.

This type of presentation is basically true in that some environments are almost totally oxidizing (e.g., nitric acid in its attack on copper) and some almost totally reducing (e.g., dilute hydrochloric acid in corrosion of zinc). On the other hand, all electrochemical corrosion is oxidizing in the sense that the anodic reaction is the dissolution of the metal, with the formation of metal ions (positively charged) through the loss of electrons.[1] There must be a corresponding cathodic reaction, a reduction process.

Actually, when we speak of a reducing acid, we mean that the more usual cathodic reaction is the reduction of hydrogen ions to atomic hydrogen, which then usually dimerizes to the molecular hydrogen which is often seen to be evolved by corrosion in dilute acid.

In an oxidizing acid, on the other hand, the cathodic reaction is the reduction of the anion (e.g., the evolution of brown oxides of nitrogen from corrosion of certain metals by nitric acid).

We must also distinguish between corrosive environments which are inherently oxidizing or reducing, and those which become so by virtue of extraneous species (e.g., oxidizing inhibitors like chromates, oxidizing contaminants like ferric ions in HCl). Acids, alkalies, and various aqueous or organic mixtures may be oxidizing or reducing primarily by virtue of such (perhaps inadvertent) additions. We shall

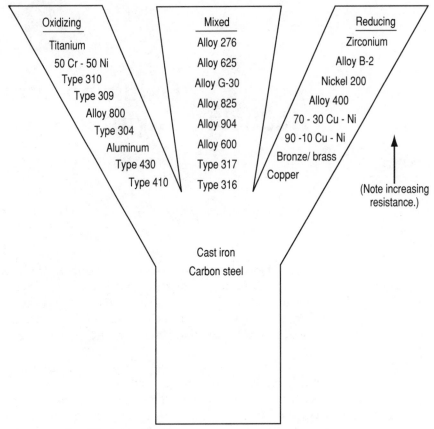

Figure 3.1   The "Y" of corrosion.

also note further below that an environment may be oxidizing to one metal or alloy and reducing to another.

Oxidizing Agents

When dealing with aqueous solutions other than specific acids or alkalis, and particularly in the pH range from about 4.5–5 through 9–9.5, the most common oxidizing agent is dissolved oxygen (DO).

It should be noted that, in listing oxidizing agents by order of potency, every oxidant is *reducing* to the next higher (stronger) oxidant. For example, nitrites are oxidizing compared to DO but are themselves oxidized by peroxides, the latter being reduced in the interaction.

Oxidants are things capable of reduction in the environment in question, in the process of oxidizing something else. Common oxidizing contaminants or inhibitors (i.e., constituents deliberately added to reduce corrosion, particularly of steel and iron-based alloys) include:

Dissolved sulfur (DS)

Dissolved oxygen

Nitrites

Nitrates

Peroxides

Chromates

Vanadates

Metallic cations of high valence state (e.g., Fe^{+++}, Cu^{++})

Most constituents commonly described as oxidizing species are conventionally thought of as stronger than dissolved oxygen, and can often be quantified by an iodimetric-type wet chemical analysis.

Reducing Agents

Reducing agents can also be listed in order of potency. The proton (i.e., H^+), which is always present in water or aqueous solutions as well as in acids, may function as an oxidizing agent (being reduced to nascent atomic hydrogen) or as a reducing agent (oxidized to the hydroxyl ion OH^- or to water). Molecular hydrogen, in common experience, is primarily a reducing agent, being oxidized to water in the reduction of ferric compounds to ferrous salts, as in acid cleaning of steel equipment, for example.

Common reducing agents are:

Hydrogen

Hydrogen sulfide

Sulfur dioxide

Hydrazine

Metallic cations of low valence state (Fe^{++}, Sn^{++}, Cu^+)

Mixtures

Solutions can be ranked in terms of redox potential, which is the net result of all possible oxidizing and/or reducing reactions which can occur on a nonreactive surface, such as a platinized titanium electrode. The redox potential is measured against a standard half-cell electrode.

Even this may not be definitive, because it fails to consider the metals or alloys of interest in the particular environment.
Nevertheless, we can make some generalizations.

1. Inherently oxidizing environments such as nitric acid become less so in the presence of reducing ions, such as fluorides or other halides. This is why such alloys as alloy C276 (UNS N10276) may be required in lieu of type 304L (UNS S30403) in nitric/HF solutions.

2. Inherently reducing solutions such as dilute hydrochloric acid become oxidizing upon contamination with oxidizing species (e.g., chromates, nitrites, ferric ions). Thus alloy B2 (UNS N10665), which will withstand HCl to the boiling point, must be replaced with the chromium-bearing alloy C276 (N10276) if there is more than a few parts per million ferric ion contamination. Even then, the permissible temperature will be much lower than the boiling point.

3. The usefulness of corrosion-resistant alloys which depend on a passive film for their resistance may fluctuate in an unpredictable manner in environments such as moderate concentrations of sulfuric or phosphoric acid when the acids are contaminated with admixtures of oxidizing (Fe^{+++}, Cu^{++}, nitrates) and reducing (fluorides, chlorides) species.

This problem of contaminants, either in supposedly pure chemicals or in process reactions, will often negate the published corrosion data or materials selection information. The data (or information) is rarely wrong, but may not be relevant to the situation at hand.

Material Considerations

It is generally true, as indicated in the literature, that metals such as copper and nickel are attacked by oxidizing environments while passivating alloys (e.g., aluminum, titanium, stainless steels, chromium-rich alloys) tend to be resistant under these conditions.

Nevertheless, in a particular situation, the environment may be oxidizing to one metal or alloy and reducing to another. Thus a 60% sulfuric acid solution at 80°C (175°F) will act as an oxidizing acid toward alloy C276 or zirconium but as a reducing environment to type 316L stainless steel or titanium. As noted above, extraneous contaminating species (e.g., Cu^{++}, chlorides) can profoundly alter this environment's behavior.

References

1. C. P. Dillon, *Corrosion Control in the Chemical Process Industries*, McGraw-Hill, New York, 1986.

Candidate Materials— Metallic

The possible materials of construction for chemical process industries range from ferrous alloys (e.g., irons and steels, with or without coatings or linings) through a variety of stainless steels, copper alloys, other nonferrous alloys (e.g., aluminum, lead, tin), nickel-rich or nickel-base alloys, to reactive metals (e.g., titanium, zirconium, tantalum) and even the precious metals themselves (e.g., silver, gold, platinum). Organic materials (e.g., plastics or elastomers) are most often employed as coatings or linings, but may be used as solid or reinforced materials where conditions of temperature and pressure permit. Inorganic materials (e.g., ceramics, carbon, graphite) are employed both as solid materials of construction and as linings (e.g., of acid brick). Nonmetallic materials are discussed in Chap. 5.

Following is a brief summary of the more important characteristics of the various materials. Table 4.1 lists materials, either by specific material or by groups of approximately equivalent materials, which will be used further in the text to indicate selections in the simplified flow diagrams. The reader will find further information in brief in Ref. 1 and in more detail in Ref. 2 and other references. [*Note:* The nomenclature used will refer to metals and alloys by their common name or alloy designation, followed by the Unified Numbering System (UNS) number in parentheses.]

Aluminum Alloys

The aluminum alloys employed in chemical processes, notably alloys 3003 (UNS A93003), 5154 (A95154), and 6061 (A96061), are approximately equivalent in chemical resistance and differ primarily in

TABLE 4.1   List of Materials

Material	Example
Aluminum alloys	
Commercially pure aluminum	A91100
Manganese alloy	A93003
Magnesium alloy	A95154
Magnesium-silicon alloy	A96063
Iron and steel	
Carbon or low-alloy steel	G1030, G4340
Gray cast or ductile iron	F10001
High-silicon cast iron	F47003
High-nickel cast irons	
15% nickel cast iron	F41000
20% nickel cast iron	F41002
30% nickel cast iron	F41004
Ductile nickel cast iron	F43000
Alloy steels	
1.5 chromium-.5 Mo	K22094
2.5 chromium-1.0 Mo	K21390
5 chromium	S50100
7 chromium	S50300
9 chromium	S50400
3.5 nickel	K32025
Stainless steels	
Martensitic	S41000
Ferritic	S43000
Austenitic (18-8)	S30403
Austenitic (18-12-3)	S31603
High-performance stainless steels	
Ferritic (1 Mo)	S44625
Ferritic (4 Mo)	S44700
Austenitic without Mo	S31003
Austenitic with Mo	S31254
Duplex	S31803
High-silicon alloys	Proprietary
Nickel-rich high-performance alloys	
Without molybdenum	N08800
3% molybdenum	N08825, N08020
6% molybdenum	N08028, N08026, N06007
Nickel-base alloys, chromium-free	
Nickel	N02200
Nickel-copper	N04400
Nickel-molybdenum	N10665
Nickel-base alloys with chromium	
Molybdenum-free	N06600
High molybdenum	N06625, N10276, N06022
Copper-base alloys	
Copper	C11000
Admiralty metals	C44300

TABLE 4.1   **List of Materials** (Continued)

Material	Example
Copper-base alloys (Continued)	
Aluminum bronze	C61400
90-10 cupronickel	C70600
70-30 cupronickel	C71500
Lead	
Common lead	L50045
Chemical lead	L51120
Antimonial lead	L52605, L53125
Tellurium lead	L51123
Silver	
Fine silver	P07016
Sterling silver	P07931
Noble metals	
Gold	P00020
Platinum	P04980
Reactive and refractory metals	
Titanium, grade 1	R50250
Titanium, grade 2	R50400
Titanium, grade 3	R50550
Titanium, grade 7Pd	R52400
Titanium, grade 11Pd	R52250
Titanium, grade 12	R53400
Zirconium, unalloyed	R60701, R60702
Zirconium-tin alloy	R60704
Zirconium-columbium alloy	R60705
Tantalum	R05210
Ceramic materials	
Glass	
Porcelain	
Chemical stoneware	
Concrete	
Brick	
Carbonaceous materials	
Carbon	
Graphite (phenolic-sealed)	
Graphite (epoxy-sealed)	
Thermoplastics	
Polyethylene	
Polypropylene	
Acrylonitrile-butadiene-styrene (ABS)	
Cellulose acetate-butyrate (CAB)	
Polyvinyl chloride (PVC)	
Chlorinated PVC (CPVC)	
Polyvinylidene chloride (PVDC)	
Polyvinylidene fluoride (PVDF)	
Fluorinated ethylene-propylene (FEP)	
Polytetrafluorethylene (PTFE)	
Perfluoralkoxy (PFA)	

TABLE 4.1 List of Materials (*Continued*)

Material	Example
Thermosetting resins	
Epoxy	
Phenolic	
Polyesters (maleic, isophthalic)	
Polyesters (chlorendic)	
Polyesters (bisphenol)	
Polyesters (brominated)	
Vinyl esters	
Furane	
Elastomers	
Natural rubber	
Butadiene-styrene (Buna S)	
Butadiene-acrylonitrile (Buna N)	
Butyl	
Neoprene	
Ethylene-propylene diene monomer	
Silicone rubber	
Chlorsulfonated polyethylene (Hypalon)	
Kel-F	
Viton	
Kalrez	
Plate	
Chromium	
Nickel	
Electroless nickel	

strength. Castings are indicated by a similar designation using 0 in lieu of 9, e.g., A0*NNNN*.

The aluminum alloys are resistant to many oxidizing acids and other chemicals, within specific parameters of use. They are resistant to organic chemicals other than chlorinated solvents and anhydrous alcohols. They resist amines in concentrated form, but are attacked by aqueous solutions as well as by other strong alkalies (e.g., caustic solutions), because of their amphoteric nature.

Pitfalls in applications of aluminum alloys include a susceptibility to pitting and concentration cell attack and, because of their anodic position toward other metals, to cementation attack by heavy metal salts in solution (e.g., Hg^{++}, Cu^{++}).

Cast Irons

Gray cast irons

Heavy-walled pipe and vessels may be constructed of ordinary (i.e., gray) cast iron, which will tolerate relatively high rates of attack.

Some pipe and most valves and pumps are preferred to be cast as the ductile iron for improved resistance to mechanical shock.

Gray cast iron may suffer graphitic corrosion in acidic environments, the iron being leached out with no apparent metal loss. This leaves a very weak structure, deceptively intact but lacking mechanical strength.

Nickel cast irons

There are several grades of austenitic nickel cast iron (from 14% to 38% nickel), which offer greatly improved corrosion resistance in some specific environments, as compared with gray iron.[3] The 20% D-2 alloy (F43000) is widely used in seawater applications, for example.

Silicon cast irons (F47003)

Silicon is added to cast irons in the range from 11 to 14.5% to produce very corrosion resistant alloys. The initial corrosion leaves a protective siliceous film which is highly protective in most environments other than caustic or hydrofluoric acid. Unfortunately, these castings are brittle and difficult to machine and weld.

The 14.5% grade (F47003) is often preferred (ASTM A-518, grade 1). Variants include grade 2, with about 4% chromium for increased resistance in chloride-bearing environments, and several proprietary grades (e.g., vacuum-treated for higher strength, with up to 16% silicon).

Silicon cast irons must be protected against thermal and/or mechanical shock, because of their low ductility and mechanical strength.

Steels

Carbon steels (e.g., SAE 1030; G10300) are used for many items of process equipment, boldly exposed in mildly corrosive situations or protected from direct attack while providing mechanical strength to the item (e.g., coated, lined, or clad construction). Low-alloy steels (e.g., SAE 4340; G43400) are used for greater strength (with substantially the same corrosion resistance as carbon steels). The high-alloy steels (2½ Cr-1 Mo or K21390 and 3½ Ni steel or K32025) are used for improved performance under high-temperature and low-temperature conditions, respectively.

Stainless Steels

The conventional stainless steels are available in three distinct groups: martensitic, ferritic, and austenitic. Except for some use of type 430 (S43000) in nitric acid, only the austenitic grades are commonly used in chemical processes (high-performance stainless steels, including ferritic varieties, and some nickel-rich alloys are discussed in "High-Performance Stainless Steels" and "Nickel-Rich High-Performance Alloys" below).

The conventional austenitic grades are exemplified primarily by types 304, 316, and 317 and their low-carbon variants. Type 304 (S30400; 18 Cr-8 Ni-.08 C) is used under oxidizing conditions (e.g., in chemicals, pharmaceuticals, food-grade products) where intergranular corrosion is not a problem but general corrosion resistance or product purity or stability might be of concern. Stabilized grades (e.g., S32100, S34700) or low-carbon grades (e.g., S30403) are employed where intergranular corrosion after heat treatment or welding is possible.[4]

Addition of a few percent molybdenum will produce the type 316 (16 Cr-12 Ni-3 Mo) and 317 alloys and their low-carbon variants and stabilized versions (type 316Ti; S31635). These have improved resistance to some chloride-bearing media as well as specific chemicals (e.g., organic acids, amines).

High-Performance Stainless Steels

The high-performance stainless steels are ferrous alloys (i.e., containing more than 50% iron), as distinguished from the nickel-rich alloys discussed in "Nickel-Rich High-Performance Alloys" below. They fall into three categories: ferritic, austenitic, and duplex (i.e., part ferrite, part austenite in about equal proportions).

Ferritic grades

The high-performance ferritics are derived from type 446 (S44600), a heat-resisting steel. Originally, vacuum-arc melting was used to produce a low-interstitial alloy, 26-1 with 1% molybdenum and less than .01% carbon, XM-27 (S44625), having superior resistance to chloride pitting and stress-corrosion cracking. Newer metallurgical techniques have produced a titanium-stabilized grade of higher carbon content, XM-33 (S44626). Higher-molybdenum-content alloys were developed for seawater applications, such as 28-4 (S44700), later with small additions of nickel, 28-4-2 (S44800). Currently, these are exemplified by proprietary alloys, such as Monit (S44635) and Seacure (S44660).

Such materials have good corrosion resistance in chloride environments, but are useful primarily in the form of heat-exchanger tubing. A really poor nil-ductility transition temperature makes them unsuitable for items of appreciable wall thickness.

Austenitic grades

Historically, these comprised alloys like types 309 (25-12, S30900) and 310 (25-20, S31000), which are primarily heat-resistant grades. Type 310S (S31008), however, is used for condenser tubing in hot nitric acid.

The primary candidate in this category today is alloy 254SMO (S31254), a 20 Cr-18 Ni-6 Mo low-carbon grade with outstanding resistance to chloride pitting and stress-corrosion cracking (SCC). (Competitive products of less than 50% Fe content are not true stainless steels; they are discussed in "Nickel-Rich High-Performance Alloys" below.)

Duplex alloys

These comprise those alloys in which the nickel content is held down to produce an approximately 50:50 distribution of ferrite and austenite, designed to improve strength (as compared with austenitic grades) as well as greater resistance to chloride effects.

Type 329 (S32900; 26 Cr-4 Ni-2 Mo) had a high carbon content (.20 C max.), which made it unsuitable for welded construction. A low-carbon variant, S32950 (.03 C max.) is now available. An early proprietary replacement, alloy 3RE60 (S31500; 18 Cr-5 Ni-3 Mo), found useful application in many services. Currently, the duplex alloys are perhaps better exemplified by alloy 2205 (S31803; 22 Cr-6 Ni-3 Mo) and alloy 255 (S32550; 26 Cr-6 Ni-3 Mo-2 Cu).

Although they have higher strength than austenitic grades of similar chloride resistance, there are potential problems in weldability and with embrittlement due either to nil-ductility transition temperature (NDTT) characteristics or to transformation of ferrite to sigma phase during welding or heat treatment. A new alloy, alloy 2507 or S32750, available as seamless tube and pipe, is alleged to be a substantial improvement, with NDTT below -50°C (-58°F), in thicknesses to 20 mm (0.79 in).

The coefficient of thermal expansion of the duplex grades is about midway between steel and the 18-8 stainless steels.

Nickel-Rich High-Performance Alloys

These comprise a group of alloys neither iron-based nor nickel-based, in the sense that neither element constitutes more than 50% of the chemical composition.

Molybdenum-free

The original molybdenum-free alloy was alloy 800 (N08800), 22 Cr-42 Fe-35 Ni-.1 C-Ti, now used primarily in heat-resistant variants.

Where molybdenum is detrimental to corrosion resistance (e.g., in oxidizing acids), it may be used for resistance to chloride pitting or SCC.

Molybdenum-bearing

The original molybdenum-bearing alloy in this category was alloy 20, later alloy 20Cb3 (N08020), approximately 20 Cr-33 Fe-36 Ni-4 Cu-3 Mo-.07 C-Cb, developed for resistance to intermediate strengths of sulfuric acid. Alloy 20Cb3 is fairly resistant to chloride pitting and SCC, and was originally the next step above types 316 and 317 when these phenomena were a problem. Alloy 825 (N08825), 22 Cr-44 Ni-26 Fe-3 Cu-3 Mo-Ti, has similar resistance.

More recently, there have been developed a group of alloys of lower nickel content (hence lower price) and *better* chloride resistance by virtue of either a higher molybdenum and copper content or of 6% molybdenum content. These are exemplified by such materials as alloy 6XN (N08367; 20 Cr-25 Ni-6 Mo-N), alloy 28 (N08028; 27 Cr-32 Ni-32 Fe-4 Mo-1 Cu) and alloy 904 (N08904; 22 Cr-26 Ni-45 Fe-4 Mo-2 Cu).

The austenitic grades have good NDTT properties and weldability, and their coefficient of thermal expansion is close to that of the 18-8 stainless steels, which is helpful in replacing heat-exchanger tubing and other maintenance operations.

Somewhat more highly alloyed is alloy G (N06007; 23 Cr-48 Ni-20 Fe-7 Mo-2 Cu-.05 C) and its variants, alloy G-3 (N06985) and alloy G-30 (N06030), used primarily for wet-process phosphoric acid applications (with sulfuric acid, Fe^{+++}, chloride, and fluoride contaminants).

Nickel-Base Alloys

The nickel-base alloys are those compositions with more than 50% nickel. The two major groups are chromium-free alloys and chromium-bearing alloys. Each group contains alloys with and without molybdenum as a further alloying element.

Chromium-free grades

Characteristically, chromium-free alloys are attacked by oxidizing environments and show useful resistance only under reducing conditions.

Alloy 200 (N02200).   This is commercially pure nickel, used primarily in strong, hot caustic solutions. There is a low-carbon variant, alloy 201 (N02201; .02 C max), for service above about 300°C (570°F). Nickel is also used as electroplating or as electroless plate (essentially a nickel-phosphorus alloy) to protect product purity.

Alloy 400 (N04400). This is the traditional Monel formulation, approximately 67 Ni-30 Cu. Resistant to both acids and alkalis in the absence of oxidizing agents, inherent or otherwise, it is subject to autocatalytic corrosion if the corrosion products (i.e., cupric ions) are allowed to concentrate in the environment.

Alloy B-2 (N10665). Originally, the Ni-30 Mo alloy was developed to resist boiling hydrochloric acid. The current alloy is of controlled chemistry to prevent sensitization and intergranular corrosion when welded or heat-treated.

It must be noted that even small amounts of oxidizing contaminants can cause corrosion in otherwise reducing acids (e.g., sulfuric, hydrochloric, phosphoric).

Chromium-bearing grades

Alloy 600 (N06600). The traditional Inconel alloy is about 16 Cr-8 Fe-.15 C nickel alloy, with excellent corrosion resistance to hot alkalies. It is also used to resist halogens and halogen acids as in high-temperature-gas (*not* normal immersion) applications.

Alloy 625 (N06625). This is derived from alloy 600 by the addition of more chromium plus molybdenum, 22 Cr-39 Ni-9 Mo-5 Fe-4 Cb-.1 C. Resistant to seawater, alloy 625 has resistance close to that of the Ni-Cr-Mo alloys [see "Alloy C276 (N10276)" below] in many corrosive services.

Alloy C276 (N10276). Historically, alloy C was developed by substituting chromium for half the molybdenum in the original alloy B, for improved resistance under oxidizing conditions. The modern version, alloy C276, and its variants alloy C22 (N06022) and C4 (N06455), resist intergranular corrosion while withstanding a wide variety of oxidizing and reducing acid environments.

Copper and Its Alloys

From the standpoint of process corrosion, we are concerned only with copper and aluminum bronze in the wrought form and a few tin bronze castings for valves and pumps. (The zinc-rich brasses and the cupronickels are selected to resist *waterside* corrosion only, e.g., Admiralty B, aluminum brass, 90-10 cupronickel.)

The copper alloys of interest resist reducing acids, but are subject to attack by both acids and alkalis if oxidizing in nature (or containing oxidants), and are subject to stress-corrosion cracking by ammonia under some conditions. They suffer general corrosion by ammonia and amines in the simultaneous presence of NH_3 or RNH_2, water, and air or oxidants, with formation of royal-blue complex corrosion products.

Lead

Lead is an amphoteric metal, subject to chemical attack by both acids and alkalis. However, it has substantial corrosion resistance in those environments which produce an insoluble film of corrosion products (e.g., sulfuric acid up to 70% at 120°C; 80% at 50°C). Common lead (L50045) has better, harder, and stronger variants, such as chemical lead (L51120), antimonial lead (L52605), and tellurium lead (L51123).

There are health and environmental problems associated with fabrication and joining of lead, and it is less widely used in modern industrial countries than previously, while still used in Third World countries particularly.

Noble Metals

Because of cost considerations, such noble metals as silver, fine (P07016) or sterling (P07931), gold (P00020), and platinum (P04980) are used only in thin sections or as electroplate where less expensive metals will not do.

Reactive Metals

The three common reactive metals (so called because of their ready combination with hydrogen, oxygen, or nitrogen at elevated temperatures) are titanium, zirconium, and tantalum.

Titanium

Titanium is used in such strongly oxidizing environments as *wet* chlorine and strong nitric acid (within certain limits) and for resistance to high-chloride environments. Note that titanium will *burn* in dry chlorine, and is susceptible to crevice corrosion in moist chlorine. It is *not* resistant to reducing acids unless they are heavily contaminated with oxidizing species (e.g., ferric or cupric ions or nitrites).

Zirconium

Zirconium is useful in intermediate strengths of sulfuric acid (approximately 60% or less, to be conservative) and to hot hydrochloric acid, in the absence of ferric ions. Corrosion, when it does occur, may produce pyrophoric products which are a possible ignition hazard.[5]

Tantalum

Tantalum has corrosion resistance similar to that of glass, being attacked by caustic and by hydrofluoric acid. It is, however, also corroded by fuming sulfuric acid, chlorosulfonic acid, and sulfur trioxide.

It is used in bayonet heaters, in thin-walled heat exchangers, as patches for repair of glass-lined equipment, and as a uniquely pore-free electroplate on less resistant substrate materials.

References

1. C. P. Dillon, *Corrosion Control in the Chemical Process Industries*, McGraw-Hill, New York, 1986.
2. *Metals Handbook*, 9th ed., vol. 13 *Corrosion*, ASM International, Metals Park, Ohio, 1987.
3. "Ni-Resists and Ductile Ni-Resists—Engineering Properties and Applications," Publication no. 1231, Nickel Development Institute, Toronto, 1987.
4. *Corrosion Handbook No. 1*, "The Forms of Corrosion-Recognition and Prevention," NACE, Houston, 1982.
5. "Pyrophoric Surfaces on Zirconium Equipment; A Potential Ignition Hazard," special technical report, Materials Technology Institute of the Chemical Process Industries, St. Louis, 1985.

5

Candidate Materials— Nonmetallic

The possible nonmetallic materials of construction for chemical process industries comprise three groups: ceramic, carbonaceous, and organic. The organic materials are the plastics and elastomers and wood.

Ceramic Materials

The term *ceramic* includes a broad spectrum of materials, including siliceous materials (cement, glass, stoneware, porcelain) and other materials (oxides, carbides, nitrides, etc.). Characteristically, ceramics can be formed into useful shapes and acquire permanence by firing at elevated temperatures.

Glass

There are three types of glass equipment utilized in the chemical industry. These are borosilicate glass (e.g., Pyrex[tm]), glass-lined steel, and "crystallized glass" (e.g., Pyroceram[tm]).

Borosilicate glass is used in pipe, heat exchangers, and columns up to about 40 in (1000 mm) in diameter. Glass-lined equipment is available as tanks, reactors, piping, valves, and even vessel agitators and pump impellers. The normal permissible temperature limit is about 175°C (350°F). Crystallized glass is used as a coating where resistance to severe thermal shock is required.

Glass is subject to damage by mechanical shock. It is resistant to common chemicals except caustic, fluorides, hot phosphoric acid, and aqueous environments above the atmospheric boiling point.

Acid brick

Acid brick is approximately 70% silica and 23% alumina plus oxides of magnesium, titanium, and alkaline metals. Silica brick (i.e., 98% silica minimum) is superior in hot acids but has no alkali resistance. Usually, brick linings in vessels should incorporate a protective membrane of plastic or rubber behind the brick to protect the steel substrate. The *mortar* must be carefully selected to meet the service conditions.

Stoneware and porcelain

Chemical stoneware is chemically equivalent to acid brick; it is usually gray or buff in color and glazed to improve resistance to wetting. Porcelain is a less porous and more homogenous material, usually white. Conventional stoneware and porcelain have the same resistance as glass, although proprietary grades (e.g., Body 97^{tm}) can be procured to resist up to 20% caustic at 100°C (212°F).

Refractories

The bulk of the refractories (i.e., heat-resistant materials) used in the chemical industry are castables, consisting of an aggregate (e.g., brick, shale, silicon carbide, alumina) in a binder (e.g., calcium aluminate, silicates, phosphates), sometimes incorporating stainless steel or other alloy fibers.

Chemical resistance depends on the specific components (e.g., silica refractories resist acid conditions but not alkalies).

Carbon and Graphite

Commercial products consist of bonded particles of carbon; articles produced below 1400°C (2550°F) are carbon while those heated over 1980°C (3600°F) have the crystalline structure of graphite.

Carbon is used primarily for brick and for tower packing. Graphite is used as flexible braid and sheet for gasketing and, because of high thermal conductivity, for heat exchangers. Fibers are used for high-strength composites with thermosetting resins.

Carbon and graphite resist acids and alkalis, except powerful oxidizing agents (e.g., nitric acid, hot sulfuric above 70%).

Organic Materials

For convenience, we will list separately the two types of "plastics" (i.e., thermoplastic and thermosetting materials). This section also covers elastomers and wood. Only materials usually utilized in the chemical process industries (CPI) are listed.

Thermoplastics

Polyethylene. As either cross-linked or ultrahigh-molecular-weight polymer, polyethylene is used as drums, tank liners, and piping. Polyethylene is subject to chemical attack by aromatic, aliphatic, and chlorinated solvents, and to stress-corrosion cracking by detergents, wetting agents, and some solvents (e.g., ketones, esters).[1]

Polypropylene. Similar to but stronger and more chemically resistant than polyethylene, polypropylene (PP) is used in dilute acids, corrosive waters, and up to 90% sulfuric acid. Reportedly, it may suffer chemical stress cracking (CSC) in dilute sulfuric acid.[2] It is most often used in lined steel pipe and as solid pipe or sheet material. A copolymer of PP and butylene is used for some plastic piping for water services, for example.

Polyvinyl chloride. Polyvinyl chloride (PVC) and its variants (e.g., chlorinated polyvinylchloride, CPVC, for somewhat higher temperature) are used for oxidizing aqueous streams (e.g., chlorine solutions) and dilute acids. They are attacked by many organic solvents. The temperature limitation is about 80°C (175°F), above which temperature even CPVC is greatly weakened.

Fluorinated plastics. These comprise a family of products of decreasing temperature capability from the fully fluorinated ethylene polymer, all having unusually good chemical resistance.

Tetrafluorethylene (PTFE). Almost immune to chemical attack except by fluorine, molten alkali, and a few exotic chlorinated solvents, PTFE has a temperature limit of about 275°C (525°F). It is used as a vessel liner, as lined pipe and valves, and as valve seats, heat exchangers, flexible hose, and envelope-type gaskets.

Perfluoralkoxy (PFA). PFA can be used to about 260°C (500°F) and, because it is melt-processible, can be molded, fabricated into vessel linings, or used as a coating.

Fluorinated ethylene-propylene (FEP). Up to its limit of about 205°C (400°F), FEP has the same chemical resistance as PTFE or PFA. It can also be molded and used for liners for vessels, pipes, and valves.

Polyvinylidene fluoride (PVDF). PVDF (e.g., Kynar^tm) has working properties similar to PVC, but much greater chemical resistance and a temperature limitation of about 150°C (300°F). [*Note:* Specialty copolymers, such as ethylene-chlortrifluorethylene (ECTFE) and ethylene-trifluoroethylene (ETFE) are beyond the scope of this book. See Ref. 3 and 4 or manufacturer's literature for detailed information.]

Thermosetting resins

There are basically four types of thermosetting resins used in the chemical process industries, usually reinforced with fiberglass or other filler materials.

Epoxies. Epoxy laminates have superior chemical resistance, especially in hot alkalies, nonoxidizing mineral acids to about 20%, and organic solvents other than amines and chlorinated solvents.

Phenolics. Heat-cured phenol-formaldehyde resins resist solvents and acids but are attacked by alkalies, oxidizing agents, and amines.

Polyesters. The common isophthalate polyesters are used primarily for aggressive waters, while vinyl esters have enhanced oxidation resistance. For chemical service, the bisphenol-A polyesters are used (e.g., in acetic acid service).

Furanes. Furfuryl-alcohol-based resins give thermosets of superior chemical resistance as compared with polyesters, resisting most types of organic and inorganic chemicals except oxidizing media and certain nitrogenous and chlorinated solvents.

Rubber and elastomers

Both natural and synthetic rubbers (elastomers) are used in chemical process equipment. They may be either boldly exposed or used as a membrane for brick-lined equipment.

Natural rubbers. Despite poor resistance in many organic chemicals and against oxidizing agents, natural rubber has been widely used in nonoxidizing acids (e.g., HCl) and alkalies.

Neoprene. The chloroprene polymer has better oxidation resistance than natural rubber and has been used as a tank lining.

Ethylene-propylene diene monomer (EPDM). The copolymer of ethylene and propylene has superior resistance to hot water and steam and has a broad spectrum of chemical resistance as well.

Specialty elastomers

1. Silicone rubbers will withstand temperatures as low as $-75°C$ ($-103°F$) and as high as 200°C (390°F).
2. Chlorsulfonated polyethylene has resistance similar to neoprene but will also withstand oxidizing environments.
3. Fluorinated elastomers will withstand oxidants and temperatures as high as 300°C (570°F), and resist acids and alkalis. They have poor solvent resistance, except for the perfluorelastomer Kalrez[tm]

Wood

Wooden tanks and piping are often useful for acid salt solutions, and the natural resistance can be enhanced by impregnation with suitable plastics. Wood should not be used with strong acids or oxidizing agents, and professional advice should be obtained for selection in corrosive or abrasive services.

References

1. C. P. Dillon, *Corrosion Control in the Chemical Process Industries*, McGraw-Hill, New York, 1986.
2. "Process Industries Corrosion—The Theory and Practice," NACE, Houston, 1986.
3. *Plastics Annual*, Society of the Plastics Industry, New York.
4. J. H. Mallinson, *Chemical Plant Design with Reinforced Plastics*, McGraw-Hill, New York, 1969.

Heavy Chemicals

In this section, the manufacture of specific heavy chemicals is described.
The general format for each chapter covers:

- *Introductory comments*

- *The process flow diagram and descriptive text*

- *Discussion of relevant materials (Note:* to avoid needless repetition, inappropriate materials of construction may be omitted in certain chapters)

- *Pitfalls (in application of specific classes of appropriate materials in the specific process)*

- *Storage and handling (material choices thought to constitute good engineering practice for specific equipment)*

For more detailed information (e.g., on effects of impurities or problems associated with the use of a heavy chemical in other reactions or processes), the reader should consult the references.

6

Ammonia

Introduction

Ammonia is a major raw material for production of fertilizers of various kinds as well as nitric acid, nitriles, various amines, and other nitrogenous materials. A pungent gas which dissolves in water to form ammonium hydroxide, it is stored and handled as the anhydrous liquid. It is substantially noncorrosive except toward copper-base and some nickel-base alloys in the simultaneous presence of water and oxygen or oxidizing agents. It can cause stress-corrosion cracking of steel under some conditions of anhydrous storage.[1]

Process

Ammonia is produced by the catalytic reaction of three volumes of hydrogen per volume of nitrogen under high pressure (i.e., 2000 to 5000 lb/in^2 gauge; 14 to 35 MPa) at about 370°C (700°F) (Fig. 6.1). The synthesis gas is formed from hydrogen, obtained from natural gas reforming or reaction of carbon monoxide with steam and nitrogen, obtained by liquefaction of air.

The syngas is dispersed through a ferric oxide catalyst ("promoted" with traces of aluminum or potassium oxides), and the exit gas is passed to a waste-heat boiler and from there to the syngas feed preheater (interchanger). The converter product gas is recycled through the compressor before being cooled to condense the liquid ammonia. A small purge is taken to prevent accumulation of inert and unreacted gases.

Liquid ammonia is flashed to atmospheric pressure in the letdown separator to remove argon as a gaseous purge. Anhydrous liquid ammonia of high purity is transferred to low-temperature atmospheric storage.

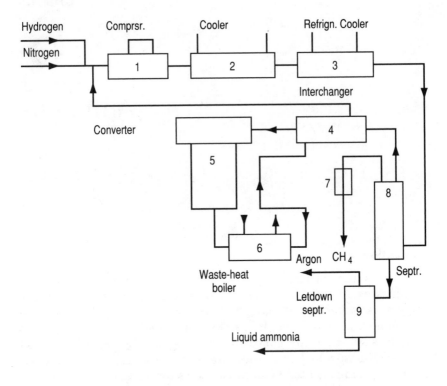

1. Carbon steel
2. Superferritic (S44625)
3. 3 1/2 % Nickel (K32025)
4. Carbon steel
5. Stainless–clad alloy steel
   (S30403/2 1/2 Cr–1 Mo; K21390)
6. K21390
7. S44625
8. K32025
9. Carbon steel

Figure 6.1 Ammonia production.

Ammonia is stored at atmospheric pressure under refrigerated conditions, i.e., *cryogenic storage* at −34°C (−29°F); at ambient temperature at moderate pressures (e.g., 300 lb/in^2 gauge; 2 MPa); or at a combination of appropriate temperatures and pressures (i.e., *semi-refrigerated storage*).

Materials

Aluminum alloys

Aluminum is resistant to anhydrous ammonia and, having good cryogenic properties, has been used as tubing in ammonia refrigeration units. It has no commercial application in the manufacturing process.

Cast iron and steel

Ammonia is noncorrosive to cast iron and steel. However, because of the potential personnel hazard, ductile iron is preferred to gray cast iron for valves. Forged-steel valves are preferred at elevated pressures, austenitic stainless steel at low temperature.

The problems with steel are (1) possible stress-corrosion cracking in ambient-temperature storage and (2) nil-ductility transition temperature embrittlement in low-temperature handling.

Stress-corrosion cracking can be alleviated by thermal stress relief at 600°C (1100°F) minimum of the storage vessel in combination with inhibition by addition of about 2000 ppm water. Note that the stress relief is a primary requirement, because water inhibition may not be effective above the liquid level as a result of the lower vapor pressure of water as compared with ammonia. Air ingress must be prevented or minimized (as little as 1 ppm of oxygen may be dangerous), and oxygen scavengers (e.g., hydrazine) or getters (e.g., internal nickel waste plates) may be helpful. High stress levels and/or "hard" welds [e.g., above Rockwell C20 or 225 Brinell hardness number (BHN)] are harmful.[2] Flame-spraying with aluminum or zinc may prevent SCC when postweld heat treatment is impractical. Nickel "getter" plates welded to the steel vessels have been suggested to remove residual oxygen necessary for SCC to occur.

$2\frac{1}{2}$ Cr-1 Mo steels are used in high-temperature service in the converter. $3\frac{1}{2}\%$ nickel steels may be employed in low-temperature service.[3]

Stainless steels

All grades of stainless steel are resistant to ammonia. Martensitic grades (e.g., type 410 or S41000) are found in turbines and compressors. Austenitic grades, perhaps up to 6% molybdenum (e.g., S31254), or superferritic grades are employed in coolers, depending on the corrosivity of the *cooling water*.

The internals of the ammonia converter are usually type 304L (S30403) or type 347 (S34700).

Copper alloys

Copper and its alloys are usually excluded from ammonia service (although resistant to the dry product) because of the possibility of stress-corrosion cracking should there be access to oxygen and water. The deep royal-blue color of corrosion products would also be objectionable.

Nickel alloys

Nickel also forms chromophoric corrosion products in the presence of air and moisture. Alloy 600 is used in such special applications as ammonia dessicators for nitriding systems and for ammonia superheaters.

Reactive and noble metals

These find no application in ammonia services, although titanium would be satisfactory for seawater-cooled exchangers. Silver is an actual hazard, forming explosive compounds (i.e., azides) with ammonia and its derivatives under some conditions.

Nonmetals

There are no practical applications of nonmetallic materials in anhydrous ammonia service, except PTFE used in packing and as the filler in spiral-wound gaskets.

Pitfalls

High-strength steel springs in inverted safety valves in ammonia tank cars are susceptible to SCC. Laboratory tests indicate that cathodic protection by aluminum alleviates the situation, and aluminized springs are now employed.

Other than the problems of ammonia cracking previously discussed, the only pitfall associated with storage in steel is corrosion under insulation resulting from ingress of moisture. This is effectively prevented by application of epoxy coatings before insulating vessels and piping.

Storage and Handling

Tanks	Stress-relieved carbon steel (70 ksi maximum tensile strength)
Tank trucks	Same
Railroad cars	Same
Piping	Stress-relieved carbon steel, 3½% Ni (K32025) steel (refrigerated)
Valves	Forged steel, ductile cast iron
Pumps	Steel, CF3M (refrigerated)
Gaskets	Spiral-wound PTFE/304

References

1. A. S. Krisher, Material Requirements for Anhydrous Ammonia, *Process Industries Corrosion—The Theory and Practice,* NACE, Houston, 1986.
2. Working party, J. M. B. Gotch, chairman, "Code of Practice for the Storage of Anhydrous Ammonia under Pressure in the United Kingdom," Chemical Industries Association, London.
3. G. Kobrin and E. S. Kopecki, "Choosing Alloys for Ammonia Services," *Materials Engineering II—Controlling Corrosion in Process Equipment,* edited by the staff of *Chemical Engineering,* McGraw-Hill, New York, 1980.

Caustic Soda

Introduction

Sodium hydroxide (caustic soda) is a coproduct of chlorine manufactured by electrolysis of a sodium chloride solution in diaphragm, membrane, or mercury electrolytic cells. The most widely used and available alkali, caustic soda ranks third in tonnage production among the heavy chemicals. It is normally produced, stored, and transported as 50% liquid, as 73% liquid, and as the anhydrous flake or solid.

Process

In a mercury cell (rubber-lined), a 25.5% sodium chloride solution is the electrolyte, which diminishes to about 21% during electrolysis. The anode may be graphite or titanium, while the cathode is a mercury-sodium amalgam. The amalgam flows to a separate compartment (the "denuder") to be hydrolyzed with demineralized water to obtain 50% caustic of exceptional purity (Fig. 7.1). Mercury cells have been discontinued because of environmental problems with mercury releases.

In the diaphragm and membrane processes, the salt solution, dissolved in rubber or brick-lined steel tanks, is subject to direct-current electrolysis in the manufacturing cell (Fig. 7.2). If naturally occurring brines are used as feedstock, they must be treated to remove calcium, iron, manganese, and sulfate impurities.

In a diaphragm cell (concrete-lined), 10 to 15% caustic is produced at the iron cathode. Across the asbestos diaphragm, graphite or titanium anodes are exposed to about 15% NaOH and 15% NaCl (sodium chloride), chlorine and hydrogen being given off. The initial caustic product is heavily contaminated with iron, sodium chlorate, sodium chloride, and dissolved chlorine.

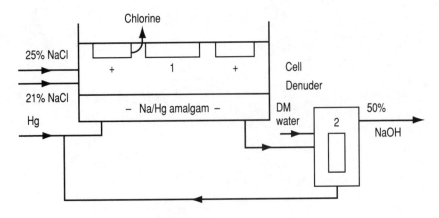

1. Carbon steel; lined (butadiene-styrene rubber)
2. Carbon electrode

Figure 7.1   Caustic by mercury cell.

The membrane cell is analogous, except that the brine feed is more highly purified and the dilute caustic product is free of the contaminants which permeate a diaphragm (because of the lower permeability of the perfluorosulfonic acid membrane).

The crude dilute caustic product from a diaphragm cell is subjected to several steps involving chemical treatment (e.g., liquid ammonia treatment of diaphragm-cell product to reduce contaminants), evaporation, and salt separation (Fig. 7.3). Evaporators are usually nickel-clad (alloy 200; UNS N02200) steel with alloy 200 heaters, or the low-carbon variant alloy 201 for temperatures above about 315°C (600°F). The life of the nickel may be limited by high chlorate concentrations (e.g., 100 ppm or more). Some manufacturers have used 26-1 (S44700) stainless steel heaters, relying on the passivating effect of chlorates to enhance corrosion resistance. Alloy 600 heaters (N06600) have been used where sulfur compounds were known or anticipated to be present.

Salt settlers are usually alloy 200. Coolers may contain alloy 400 (N04400) or alloy 625 (N06625) tubes to resist waterside corrosion, while salt settlers and slurry tanks are usually alloy 400.

Materials

Aluminum and its alloys

Aluminum alloys have no application in caustic service.

A. Diaphragm (vacuum-deposited asbestos)

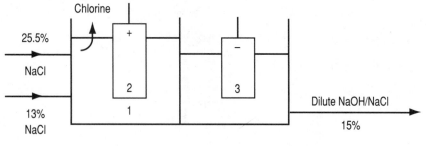

1. Brick-lined steel
2. Carbon anode
3. Steel cathode

B. Membrane (perfluorosulfuric acid)

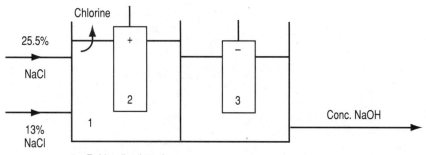

1. Rubber-lined steel
2. Titanium anode (R50250)
3. Nickel cathode (N02200)

Figure 7.2   Caustic by diaphragm and membrane cell.

Steel and cast iron

If iron contamination is not objectionable, steel and cast iron may be used in up to 70% concentration to about 80°C (175°F) *except* for potential problems with stress-corrosion cracking and the safety aspects of cast iron. For further information, see Refs. 1–3. (Although cast-iron evaporators were used for many years, gray cast iron should not be used in caustic service today; ductile cast iron or nickel cast irons are a better choice.)

Welded, cold-formed or otherwise stressed steel is subject to so-called *caustic embrittlement* (i.e., stress-corrosion cracking) in greater than 50% caustic at temperatures above about 60°C (140°F). Because

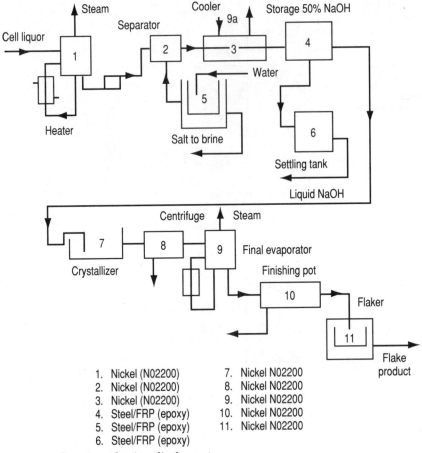

1. Nickel (N02200)
2. Nickel (N02200)
3. Nickel (N02200)
4. Steel/FRP (epoxy)
5. Steel/FRP (epoxy)
6. Steel/FRP (epoxy)

7. Nickel N02200
8. Nickel N02200
9. Nickel N02200
10. Nickel N02200
11. Nickel N02200

Figure 7.3   Caustic production (diaphragm).

concentrated sodium hydroxide must be heated, by steam tracing or internal coils, to prevent freezing at room temperature, this problem is always potentially present. Where iron contamination is not a consideration, steel piping and tanks may be used, if thermally stress-relieved after welding or field flaring.

Resistance of cast irons is greatly enhanced by nickel additions (e.g., use of austenitic 15 to 35% nickel cast irons; ASTM A436, grade 1 or F41000). Ductile or nickel cast irons are not subject to SCC.[4]

Stainless steels

Conventional martensitic and ferritic stainless steels are not used in caustic service. The "superferritics" exemplified by S44700 etc. have been used as heat-exchanger tubes, when chlorates are present in sufficient amounts to maintain passivity.

The standard 18-8 varieties (e.g., S30403, S31603) are useful up to about 80°C (175°F) in 40% caustic and to about 70°C (160°F) in 50% concentration. They may go active and suffer worse corrosion than steel above those limits and are subject to stress-corrosion cracking above about 120°C (250°F).[5] However, chloride contamination (which may reach 6000 ppm in diaphragm product) is not a problem as to either pitting or SCC of these alloys.

Copper and its alloys

There is little data on copper alloys in caustic because copper ion contamination is usually objectionable in the final product. However, copper and zincless bronzes are otherwise usefully resistant in 25 to 73% caustic, in the absence of chlorate or hypochlorite contamination.

Nickel alloys

As indicated above, nickel alloys 200 and 201 are the preferred materials for hot caustic. They may be cathodically protected to reduce nickel ion contamination. Alloy 600 may be used for heating coils and for piping (being more readily welded in the field than alloy 200) where small amounts of hexavalent chromium ion contamination can be tolerated.

The nickel-rich high-performance alloys (e.g., 6XN, 20Cb3, 800) are rarely used, unless required by cooling-water problems. The high-nickel, molybdenum-rich alloys (e.g., B-2 or N10665, C276 or N10276) are not normally employed in caustic services, for economic reasons.

Reactive metals

Titanium is resistant to up to 40% caustic at about 80°C (175°F), and its passive film may be enhanced by the presence of oxidizing contaminants. Normally, it is *less* resistant than nickel, and more expensive. It is, however, used for brine heaters preceding the electrolytic cells.

Zirconium is resistant to 50% caustic to about 60°C (140°F), but has no usual application, while tantalum is severely attacked by caustic.

Precious metals

Silver antedates nickel alloys for caustic service and is still used for concentration of strong, meticulous quality caustic (e.g., 73%).

Other metals

Lead, tin, and zinc, being amphoteric in nature, find no application in caustic service.

Nonmetallic materials

Organic. Unreinforced plastics resist caustic up to their usual temperature and pressure limitations. Fiberglass-reinforced plastics require a protective veil (e.g., of NEXUS^{tm} or Dacron^{tm}) below the gel coat.

Elastomeric sheet linings and organic coatings (e.g., neoprene latex, modified epoxy) are used to prevent corrosion of or contamination by steel, within their normal temperature limitations.

Inorganic. Carbon and graphite resist caustic, but, in practice, temperature limitations may be imposed by organic binders or cements.

Glass and ceramics are attacked by alkalies even at relatively low temperatures.

Pitfalls

Steel

Steel suffers accelerated attack, as well as premature stress-corrosion cracking, if pipe is allowed to stagnate while steam-traced; the solution temperature rises well above that needed to prevent freezing.

Stainless steels (18-8 variety)

Austenitic stainless steels may go active and suffer rapid corrosion under the same circumstances, or where steam tracing is allowed direct contact with the piping wall. At high temperatures, SCC is commonly encountered. Chlorate or hypochlorite content may profoundly affect the active/passive behavior of both austenitic and superferritic grades.

Nickel

Unexpectedly high rates may be encountered with chromium-free nickel alloys (e.g., 200, 400) if chlorate contamination exceeds about 100 ppm.

In evaporators, N06600 is preferable to N02200, particularly when steam is contaminated with carbon dioxide.

Upset conditions in mercury-cell processes can lead to liquid metal embrittlement of nickel or alloy 400, while sulfur contamination makes welding of any high-nickel alloys difficult or impossible.

Storage and Handling

Use of the following materials of construction is common for handling and storage and is thought to constitute good engineering practice:

Tanks   Carbon steel, stress-relieved (or rubber- or epoxy-coated when iron contamination is of concern).

Tank trucks   Rubber-lined or epoxy-coated steel; type 304L or 316L stainless steel. (*Note:* an alloy 825 truck has also been used for broad versatility; i.e., to handle both acids and alkalies). Nickel-clad or nickel-plated for 73% NaOH.

Railroad cars   Lined or nickel-plated steel; type 304L or 316L stainless steel.

Piping   Stress-relieved steel, PP-lined steel, type 304L or alloy 600.

Valves   Ductile iron; steel with stainless trim or CF3M.

Pumps   Ductile iron, nickel cast iron, CF3M.

Gaskets   Elastomeric, flexible graphite or spiral-wound type 304/PTFE.

References

1. J. K. Nelson, "Corrosion by Alkalis and Hypochlorite," Vol. 13, *Metals Handbook,* ASM International, Metals Park, Ohio, 1987.
2. P. J. Gegner, "Corrosion Resistance of Materials in Alkalis and Hypochlorites," *Process Industries Corrosion,* NACE, Houston, 1975.
3. J. K. Nelson, "Materials of Construction for Alkalis and Hypochlorites," *Process Industries Corrosion—The Theory and Practice,* NACE, Houston, 1986.
4. "Corrosion Resistance of Nickel and Nickel-Containing Alloys in Caustic Soda and Other Alkalis," *Corrosion Engineering Bulletin,* CEB-2; Inco Alloys International, Huntington, W. Va., 1975.
5. C. P. Dillon, *Corrosion Control in the Chemical Process Industries,* McGraw-Hill, New York, 1986.

Chlorine

Introduction

Chlorine is produced as a coproduct in the electrolytic manufacture of sodium hydroxide, as previously described in Chap. 7. Dry chlorine is a poisonous greenish-yellow gas and a powerful oxidizing agent. While the dry gas is not very corrosive below about 90°C (195°F), even traces of moisture form stoichiometric amounts of a mixture of hypochlorous acid (HOCl) and hydrochloric acid (HCl) which is severely corrosive to most metals and alloys.

While the dry product in either gaseous or liquid form can be handled in steel at ambient temperatures, some metals (e.g., titanium) will ignite and burn in dry chlorine containing less than about 3500 ppm water. Other metals and alloys (e.g., steel, cast iron, and copper-base materials) will ignite and burn in chlorine above about 205°C (400°F), although high-nickel alloys (e.g., alloys 200, 600, and B-2) may be used at about 540°C (1000°F).

Process

Wet chlorine from the anodes of the electrolytic cell is cooled and enters the base of the drying tower to be dried by a countercurrent flow of concentrated sulfuric acid (Fig. 8.1). The hot, wet cell gas may be handled in piping of reinforced thermosetting resins, borosilicate glass (i.e., Pyrex^{tm}), or stoneware. The preferred material for the cooler is titanium. The piping after the cooler may be rubber-lined steel. The drying towers are usually brick-lined steel with an elastomeric or PVC membrane. Piping between the two drying towers usually employed should be alloy C276 (N10276) or PVDF-lined steel to resist entrained sulfuric acid. The dried gas, which can now be handled in carbon-steel piping, is liquefied by refrigerant gas at cryogenic tem-

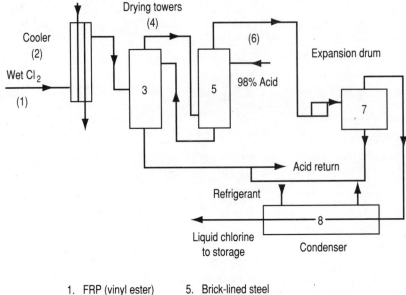

1. FRP (vinyl ester)   5. Brick-lined steel
2. Titanium (R50400)   6. Carbon steel
3. Brick-lined steel   7. Carbon steel
4. N10276   8. Carbon steel

Figure 8.1 Chlorine production.

peratures and stored under pressure in steel cylinders or tank cars. A 3% nickel steel (i.e., K31918 or equivalent) may be used for low-temperature impact strength.

Materials

Aluminum alloys

While theoretically resistant to dry chlorine to about 120°C (250°F), aluminum and its alloys are not used in practice because of possible ignition and because of severe corrosion should ingress of moisture occur.

Cast iron and steel

Cast iron is not recommended for chlorine service because of the hazard inherent in its brittle characteristics, although it is resistant to about 175°C (345°F).

Steel is the preferred material of construction for handling dry chlorine (e.g., shipment, storage, and vaporization), subject to the following warnings. Steel can ignite at elevated temperatures, as described above. The resistance of steel depends on a film of iron chlorides which can interfere with proper closure of tight-fitting parts (e.g., valve seats, valve stems). As described further below, high-nickel alloys must be used for such components, particularly where ingress of moisture might occur.

Practical temperature limits for steel vessels or piping, predicated on a corrosion rate of less than 20 mils per year (mpy), and for tubes or internal components (based on 3 mpy) are 200°C (390°F) and 150°C (300°F), respectively.

Stainless steels

Although somewhat more resistant than carbon steel to dry chlorine, stainless steel is not recommended, both because of cost and because of its propensity for chloride pitting and/or stress-corrosion cracking should ingress of moisture occur.

However, cast alloy 20 valves (CN7M; N08007) are used for valves in refrigerated chlorine equipment, where trace amounts of wet chlorine may be entrapped in rime ice.

Copper alloys

Copper and its alloys will resist dry chlorine up to about 200°C (390°F), but are subject to severe attack if ingress of moisture should occur. Flexible, annealed copper tubing is sometimes used for connections to gas cylinders, tanks, and tank cars. It must be replaced periodically because it becomes hard and rigid (from cold working) after continued use.

Nickel alloys

The non-chromium-type alloys, e.g., alloy 200 (N02200), alloy 400 (N04400), and alloy B-2 (N10665) resist dry chlorine. Recommended temperature limits for vessels and tubing are 500°C (930°F) and 410°C (770°F) for alloys 201 (N02201) and 400, respectively. Alloy 201 is required above about 425°C (800°F) to resist graphitization. Alloy 400 is used for the seats of forged steel valves.

The chromium-bearing alloy 600 (N06600) may be used as a substitute for alloy 201, and has higher strength. The alloy C276 (UNS N10276) is preferred where ingress of moisture might occur (e.g., for valve stems in valves that are exposed to atmospheric humidity).

Reactive metals

Titanium cannot be used in chlorine unless the gas contains at least 3500 ppm water. It is used primarily as an anode in electrolytic cells and as coolers for the wet cell gas. It may suffer corrosion in *crevices,* where an imbalance of HOCl/HCl may develop, even in wet chlorine.

Zirconium offers no advantage in chlorine service.

Tantalum will resist both wet and dry chlorine up to about 150°C (300°F) and is used for diaphragms and needle valves for proportioning chlorine feed to aqueous systems.

Noble metals

Gold and silver are attacked by chlorine at relatively low temperatures and find no application in chlorine service.

Nonmetals

Organic. Rubber lining of the semihard or hard variety has been widely used for dry chlorine, but is not satisfactorily resistant to wet chlorine. Traces of harmful organic solvent contaminants may be present in chlorine from reaction with carbonaceous electrodes or asphaltic-type sealants.

Polyvinyl chloride and vinyl esters (FRP) may be used up to about 90°C (195°F) in wet chlorine. Only fluorinated plastics should be used in dry chlorine (e.g., for envelope gaskets, valve diaphragms).

Inorganic. Carbon and graphite can be used in either wet or dry chlorine, but there is little occasion to do so except in gasketing. Wet chlorine attacks the binders in impervious graphite.

Glass and ceramics are impervious to chlorine attack, but are subject to thermal and mechanical shock, with the attendant hazard of possible chlorine release.

Pitfalls

The pitfalls associated with handling of chlorine in metals and alloys are primarily associated with unforeseen ingress of moisture and corrosion by the hydrochloric-hypochlorous acid mixtures formed. The possibility of burning of metals in chlorine has been previously mentioned.

Chlorine-hydrogen mixtures are exothermic and explosive. Contamination with organic materials is also dangerous.

Storage and Handling

Use of the following materials is common in the storage and handling of dry chlorine and is thought to constitute good engineering practice:

Tanks	Carbon steel (3½% nickel steel for low temperature)
Tank trucks	Carbon steel
Railroad cars	Carbon steel
Piping	Carbon steel
Valves	Carbon steel (alloy 400 seats, alloy C276 stems)
Pumps	Ductile iron, Ni-resist or alloy 400
Gaskets	Spiral-wound alloy 400/PTFE

Note. It is essential that chlorine lines be dry, organic-free, and free of corrosion-generated hydrogen.

References

1. E. T. Liening, "Corrosion by Chlorine," *Metals Handbook*, 9th ed., vol. 13 *Corrosion*, ASM International, Metals Park, Ohio, 1987, p. 1170.
2. "Process Industries Corrosion," *Metals Handbook*, vol. 13.
3. "Process Industries Corrosion—The Theory and Practice," *Metals Handbook*, vol. 13.
4. *Metals Handbook*, vol. 13.
5. "Resistance of Nickel and High Nickel Alloys to Corrosion by Hydrochloric Acid, Hydrogen Chloride and Chlorine," *Corrosion Engineering Bulletin*, CEB-3, Inco Alloys International, Huntington, W. Va., 1972.

9

Fluorine

Introduction

Fluorine is a very reactive gaseous element, but the dry product is substantially noncorrosive to metals because of the formation of insoluble corrosion product films. It rapidly oxidizes organic materials, except for the inert fluorinated plastics (e.g., PTFE). It is a severe personnel hazard if inadvertently released, forming HF and HOF with atmospheric moisture in a manner analogous to the reaction of chlorine and other halogens.

Process

Fluorine is produced by the electrolysis of anhydrous HF from molten potassium fluoride and hydrogen fluoride (KF:2HF), sometimes with a trace of lithium fluoride as an additive (Fig. 9.1).

Carbon or nickel alloy 200 (N02200) anodes are employed in a water-jacketed alloy 400 (N04400) cell. Nickel is preferred (1) for somewhat higher cell temperatures, e.g., above 200°C (390°F); (2) where anode sludge products, typical of carbon anodes, are intolerable; and (3) for better resistance to minor explosions in the anode section of the cell. Carbon anodes do have advantages in current capability, power efficiency, and electrolyte economy.

Fluorine produced at the anodes is purified, the residual HF being removed either by chilling or by absorption in sodium fluoride.

The cathodes, usually steel (although silver has been employed), are separated from the anodic section by alloy 400 screens.

The fluorine is produced under anhydrous conditions and is directly compressed by alloy 400 diaphragm pumps for liquid storage in clean, dry cylinders constructed of steel, copper, or nickel. Alloy 400 or alloy 200 cylinders offer maximum safety because they are difficult to ig-

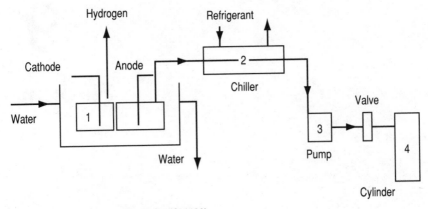

1. 70 Ni–30 Cu (N04400)
2. N04400
3. N04400
4. Carbon steel, N04400 or nickel N02200

Figure 9.1   Fluorine.

nite. (Fluorine constitutes a fire hazard when contaminated with water or grease: steel will burn in fluorine when ignited.)

Materials

Light alloys

Aluminum and magnesium resist fluorine up to 450°C (840°F) and 300°C (570°F), respectively. However, they do not normally find commercial application because of the hazard of potential ingress of moisture. In some foreign plants, magnesium cells have been employed with silver cathodes, the alloy 400 diaphragms being silver-soldered to the cell wall.

Cast iron and steel

Carbon steels with less than .01% silicon are resistant up to about 350°C (660°F), but corrosion increases with increasing silicon content, and ignition is possible in the event of dirt or grease contamination (e.g., in joints).

Cast irons are unsuitable, both because of silicon levels and the hazard inherent in brittle materials.

Stainless steels

The austenitic 18-8 stainless steels, such as type 304L (S30403), resist fluorine up to about 250°C (480°F). Resistance may be limited by the inherent silicon content, normally about 1% maximum. In addition, the columbium-bearing grade, type 347 (S34700), suffers higher attack, because of the formation of a volatile columbium pentafluoride.

However, both type 321 (S32100) and type 347 (S34700) were unaffected (less than .03 mpy) in alternate exposure to liquid and gaseous fluorine. The 18-8 stainless steels did not .ignite when tested by dynamite-cap explosion while filled with liquid fluorine.

Wet fluorine would be severely corrosive.

Copper alloys

Copper resists dry fluorine, but not at elevated temperature. Its use is limited to tubing or containers operating at ambient temperatures. Copper alloys would be rapidly attacked in the presence of moisture, because of the action of hypofluorites.

Nickel alloys

The non-chromium-bearing alloys, alloy 400 (N04400) and alloy 200 (N02200), are the preferred materials for safe handling of fluorine.[1] They are remarkably free of any ignition tendency and have excellent low-temperature impact properties. They also have much better high-temperature resistance, e.g., up to 400°C (750°F), than the chromium-bearing alloy 600 (N06600) or the nickel-bearing stainless steels. Weld rods should be of low silicon and low columbium content.

Reactive metals

Of the reactive metals, titanium resists dry fluorine (unlike chlorine) at room temperature. It is not employed, because of its extreme sensitivity to traces of HF and water.[2]

Zirconium is similar in nature, resistant to moderately elevated temperatures, but tantalum is rapidly attacked even at ambient temperature.

Noble metals

Silver is less resistant than alloy 200, and finds no application with the exception of the cathodes mentioned previously in some cell designs. Gold is resistant but platinum corrodes catastrophically.

Nonmetals

Organic. Carbon is adequately resistant to dry fluorine but forms some carbon fluorides, which may cause swelling of carbon anodes.

Breakage problems may be partly overcome by impregnation with copper.

The fluorinated plastics are resistant, but other thermoplastic and thermosetting resins (and elastomers) are rapidly attacked by oxidation.

Inorganic. The siliceous materials (e.g., glass, stoneware) are resistant up to about 150°C (in the absence of HF contamination). Alumina is resistant also.[3]

Storage and Handling

Tanks   Alloy 200 or alloy 400 preferred. Low-silicon steel permissible.

Piping   Alloy 200 or alloy 400 preferred. Copper or low-silicon steel permissible

Valves   Alloy 200 or alloy 400 with PTFE stem packing (*Note:* Provide shields and extension handles; double valving recommended.)

Pumps   Alloy 400 diaphragm pumps

Gaskets   Fluorinated plastic envelope; spiral-wound alloy 400/PTFE

References

1. "Corrosion Resistance of Nickel-Containing Alloys in Hydrofluoric Acid, Hydrogen Fluoride and Fluorine," *Corrosion Engineering Bulletin*, CEB-5, Inco Alloys International, Huntington, W. Va., 1968.
2. D. R. Hise, "Corrosion from the Halogens," *Process Industries Corrosion*, NACE, Houston, 1975.
3. E. E. Liening, "Materials of Construction for the Halogens," *Process Industries Corrosion—The Theory and Practice*, NACE, Houston, 1986.

Chapter

10

Hydrazine

Introduction

Hydrazine (N_2H_4) is a powerful reducing agent used as an oxygen scavenger in boiler feedwater treatment, as one component of rocket fuel (to combine with oxidizing agents), and for other chemical reactions. It is carcinogenic and requires adequate personnel protection against exposure.

Process

Hydrazine is produced by the partial oxidation of ammonium hydroxide with sodium hypochlorite, and the dilute hydrazine hydrate solution is dehydrated to form the final product, the monohydrate: 85% hydrazine hydrate, equivalent to 54.4% hydrazine in water (Fig. 10.1).

A 28% ammonium hydroxide solution and a 1 N sodium hypochlorite solution are mixed in a glass-lined steel tank, the mixture being fed to the intake of a cast-steel high-pressure pump with a chromium-plated plunger. The reaction mixture is heated at about 440 lb/in^2 gauge (0.625 MPa), passed through a serpentine coil for increased reaction time, then allowed to expand and cool at atmospheric pressure in two expansion vessels containing stainless-steel packing. Large amounts of unreacted ammonia are boiled off and returned to the ammonia storage tank, while the reaction product (3% hydrazine hydrate and salt) is collected, cooled, and passed to a salt evaporator and settler. The calandria on the salt evaporator contains type 316 tubes, smoothly finished and operating at a flow velocity of 12 to 15 ft/s (4.5 m/s). (*Note:* The molybdenum-bearing grades are not used elsewhere in the process; type 304 is preferred because it does not catalyze decomposition of the concentrated product.)

The monohydrate is dewatered in three successive type 304L

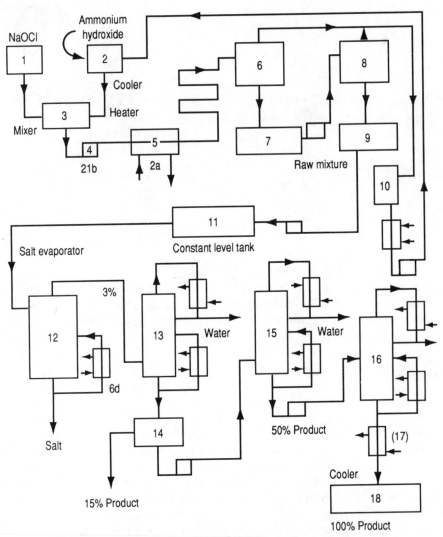

Figure 10.1 Hydrazine hydrate.

1. Carbon steel
2. Carbon steel
3. Concrete
4. Nickel–plated steel
5. Carbon steel
6. Steel with chemical stoneware
7. Carbon steel
8. Steel with chemical stoneware
9. Carbon steel
10. Carbon steel
11. Carbon steel
12. S30403
13. S30403
14. S30403
15. S30403
16. S30403
17. S30403
18. S30403

(S30403) fractionation columns, producing 15%, 50%, and 100% hydrazine hydrate respectively. A 20% solution is catalyzed with hydroquinone (for more rapid reaction with dissolved oxygen) and sold commercially for deoxygenation of boiler feedwater.

A newer PCVK process substitutes 70% hydrogen peroxide as the oxidant, in lieu of hypochlorite, eliminating the corrosion potential associated with sodium chloride and reducing contamination effects. The reaction is effected in a methyl ethyl ketone environment with a nitrile or amide and disodium phosphate as a cocatalyst. All type 304L equipment is used in this process (Fig. 10.2).

Materials

Concentrated hydrazine solutions are catalytically decomposed by contact with metallic oxide (e.g., of iron or copper) as well as by porous asbestos, activated carbon, and welding slag. Stainless or nickel alloys containing molybdenum (e.g., S31603, N10276) also catalyze decomposition.

Aluminum alloys

Aluminum is resistant to salt-free solutions of hydrazine.

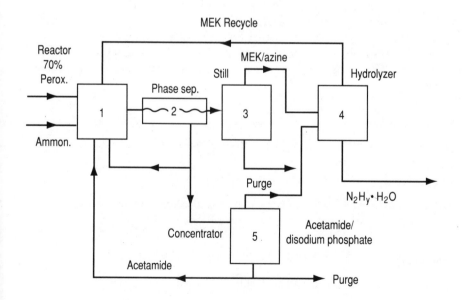

1. – 5. S30403

Figure 10.2 Hydrazine hydrate (PCVK process).

Cast iron and steel

Iron and steel show moderate corrosion rates in higher concentrations, although the 3% solution with its salt contamination can be handled in steel up to the salt evaporator (iron contamination leaving with the salt slurry).

Stainless steels

Type 304L (S30403) is resistant to all concentrations of hydrazine hydrate. The molybdenum-bearing grades have been used in those parts of the system which retain sodium chloride, provided the flow rate is high enough to prevent salt deposition. However, they must not be used with the concentrated solutions, in which type 304L or type 347 (S34700), welded by inert-gas-shielding techniques, is the preferred alloy. A limitation of less than 0.5% Mo is usually specified for the 18-8 grades of stainless steel.

Copper alloys

Copper and its alloys are not employed in hydrazine service.

Nickel alloys

The salt-contaminated hydrazine could be exposed to alloy 400 (N04400) under *anaerobic* conditions.

The preferred alloy for salt evaporation components is the molybdenum-rich, chromium-bearing alloy C276 (N10276) or alloy 625 (N06625). These are not, however, suitable for the concentrated product.

Reactive metals

There is no reported usage of reactive metals in this service.

Noble metals

Silver is attacked by hydrazine. Gold and platinum are resistant to about 100°C, but have no application in this service.

Nonmetals

Organic. Flexible graphite is a suitable gasket or packing material for hydrazine solutions.

Although PVC may be used to handle the sodium hypochlorite feed, it should not be exposed to the reaction product. Chlorsulfonated polyethylene and *butyl* rubber are the only organic plastics/elastomers listed as resistant to hydrazine, other than the fluorinated plastics (PTFE or PFA, FEP, PVDF).

Inorganic. Glass and glass-lined equipment has been used, but glass and similar ceramics are attacked by hot hydrazine solutions above about 45% concentration. The porcelain packing originally used in expansion vessels suffered enough deterioration to require replacement with stainless steel.

Pitfalls

The sodium chloride content through the salt separation step poses potential problems of chloride pitting or stress-corrosion cracking for the 18-8 stainless steels. High-molybdenum grades would be preferred, as also for brackish cooling waters, but only in the dilute solutions.

Plastics and elastomers otherwise suitable for ammonium hydroxide may suffer attack by preferential absorption of traces of hydrazine over an extended period of time.

As with other ammonia-type systems, inadvertent ingress of carbon dioxide can form carbamates which aggravate the corrosion characteristics toward ferrous materials.

Nitrogen padding in containers is necessary for acceptable stability, preventing reaction with air or oxygen.

Storage and Handling

Tanks	Aluminum, type 304 (S30400)
Tank trucks	Aluminum, type 304 (S30400)
Railroad cars	Aluminum, type 304 (S30400)
Piping	Aluminum, type 304L (S30403)
Valves	CF8
Pumps	CF8
Gaskets	Braided (packing) or flexible graphite; spiral-wound type 304/PTFE; PTFE envelope; butyl rubber

Further Reading

Audrieth, L.F., and B. A. Ogg, *The Chemistry of Hydrazine*, John Wiley and Sons, New York, 1951.
Byrkit G. D., and G. A. Michalek, "Hydrazine in Organic Chemistry," *Industrial and Engineering Chemistry*, vol. 42, 1950.
Clark, C. C., *Hydrazine*, 1st ed., (Olin) Mathieson Corporation, Baltimore, 1953.
"Hydrazine Hydrate (85%)," Bulletin HC-301, Olin Mathieson Corp., Baltimore.

"Hyzeen," Bulletin 589, Betz Laboratories, Inc., Philadelphia.
"Mathieson Hydrazine Solution, 85% Technical Monohydrate (N_2H_4 54.4% Min.)," Bulletin ML-348-158, Olin Mathieson Corp., Baltimore.
Troyan, J. E., "Properties, Production and Use of Hydrazine," *Industrial and Engineering Chemistry*, vol. 45, 1953.

Hydrochloric Acid

Introduction

Hydrochloric acid, a strong reducing acid with a pungent odor, is an aqueous solution of hydrogen chloride. It is commonly used in acid pickling or acid treatment, chemical cleaning, and chemical processing (e.g., hydrochlorination, chlorination of unsaturated organic compounds).

There are three strengths normally encountered, as follows:

Concentration, %	Sp. gr.	°Baume	°Twaddell
27.92	1.1417	18	28.34
31.45	1.1600	20	32
35.21	1.1789	22	35.78

Chemically pure (CP) acid is nominally 35.5% strength. The constant boiling mixture (CBM) is 20.2% at standard conditions; concentrated solutions tend to lose hydrogen chloride and dilute solutions water vapor to approach this concentration.

Commercial *muriatic acid* contains iron salts (Fe^{+++}) as a contaminant, while industrial "recovered" acid may be contaminated with organic chlorides or chlorine. Oxidizing agents can profoundly affect behavior of metals and alloys in HCl, while organic contaminants adversely affect plastics and elastomers.

This discussion concerns only pure HCl, unless specifically otherwise noted.

Process

The older processes, which now account for less than 4% of HCl production, involve reaction of sodium (sometimes potassium) chloride with concentrated sulfuric acid or with sulfur dioxide-air-water mixtures at elevated temperatures. However, the burning of hydrogen in chlorine is an old process still employed in production of high-purity HCl.

The old salt process fed NaCl and sulfuric acid to a fired furnace, while the Mannheim process involves a fluidized bed of salt over which sulfuric acid vapors are passed, producing 30 to 60% HCl. In the air-SO$_2$ process in the 550 to 600°C (approximately 1075°F) range, 10 to 12% product is obtained. The product is passed through acidproof cyclone separators, cooling towers, and filters to a cooler-absorber and, if necessary, to a stripper which concentrates weak HCl (Fig. 11.1).

Linings of brick, impervious graphite, rubber, or plastic (where temperatures permit) are used to handle the hydrochloric acid and its vapors.

Modern synthesis processes burn hydrogen in chlorine, yielding a high-purity hydrogen chloride for water absorption.

Unfortunately, more than 90% of hydrochloric acid as currently manufactured is recovered from organic syntheses, and may be contaminated with chlorine, organics, chloroorganics, and catalyst parti-

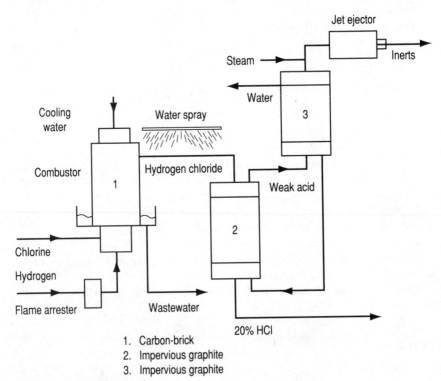

1. Carbon-brick
2. Impervious graphite
3. Impervious graphite

Figure 11.1   Hydrochloric acid.

cles unless specifically purified. Chlorine contamination may be removed by passage over activated carbon in the presence of olefins or by reaction with high-boiling paraffin hydrocarbons, while hydrogen fluoride can be removed by reaction with calcium chloride, alumina, or silica. The hydrogen chloride may be dried by passage through concentrated sulfuric acid. Chlorosulfuric acid will reduce the water content to less than 10 ppm.

Materials

Aluminum alloys

Aluminum and its alloys are totally incompatible with hydrochloric acid.

Cast iron and steel

Cast iron and steel are rapidly attacked by hydrochloric acid at all concentrations. However, cast-iron pots are still used to produce by-product HCl in the manufacture of globular bisulfated soda. It should be noted that *inhibited* 10 to 15% HCl is used in chemical cleaning and without excessive corrosion of steel when properly applied. However, cast iron is *not* protected by the inhibitor because of the galvanic influence of the contained graphite.

Stainless steels

All grades of stainless steel are rapidly attacked by HCl, changing to the active rather than passive state and corroding even faster than steel. Further, residual chlorides from such exposure can contribute to pitting or SCC in subsequent operations.

Copper alloys

Copper and its alloys are attacked by hydrochloric acid through the effects of dissolved oxygen; the accumulation of cupric ions as corrosion products makes the corrosion autocatalytic. The yellow brasses suffer dezincification in such acid conditions as well.

Nickel alloys

Nickel alloys 200 (N02200) and 400 (N04400) are not normally useful in HCl. [An exception is the use of alloy 400 in HCl-based pickling baths, where hydrogen evolution from ferrous metals generates enough free hydrogen to remove dissolved oxygen (DO) and keep metallic cations in a reduced state.]

The 30% molybdenum-nickel alloys , such as alloy B-2 (N10665), are very resistant to HCl up to the boiling point. However, even trace amounts of oxidizing contaminants (e.g., Fe^{+++}) will cause rapid attack, and it is difficult to exclude them in practice.[1]

The chromium-bearing alloy 600 (N06600) is not resistant, suffering both general corrosion and pitting. The Cr-Mo-Ni grades, such as alloy C-276 (N10276) and alloy C22 (N06022) are moderately resistant when ferric ion contamination occurs, but only in dilute solutions at relatively low temperatures.

Reactive metals

Unalloyed titanium is substantially nonresistant to pure hydrochloric acid. Even when attacked at relatively low rates in dilute acid, there is sufficient hydrogen pickup to impair mechanical properties. (Occasional successful exposure of titanium to HCl involves unrecognized contamination with ferric or cupric ions which extend the tolerance of titanium.) Ti-Pd alloys (R52400) will tolerate somewhat stronger acid than R50250 etc., in the unalloyed grades.

Zirconium (R60701, R60702) will resist HCl up to 120°C (250°F), and as high as 200°C (390°F) in 15% acid, *provided* there are not more than 50 ppm oxidizing cations (e.g., Fe^{+++}, Cu^{++}) present. A sort of "weld-decay" in the heat-affected zones may be encountered if *high-purity* zirconium is not employed, and the presence of oxidants can form pyrophoric corrosion products (see Chap. 66).

Tantalum offers useful resistance to about 175°C (345°F), but is attacked by the vapors of concentrated acid much lower than indicated in some published literature, e.g., 130°C (265°F).

Noble metals

Silver and gold will withstand HCl only at room temperature, but platinum is resistant to concentrated acid to 300°C (570°F).

Nonmetals

Nonmetallic materials are preferred for handling and storing HCl.

Organic. Rubber-lined equipment will handle HCl up to 80°C (175°F). However, the lining is sensitive to contamination by even a few parts per million of organic solvents, which are preferentially extracted and concentrated in the rubber. Aromatic solvents swell the rubber itself, while chlorinated solvents tend to dissolve the cement of the rubber-to-steel bond.

Plastics, such as PE, PP, and PVC, are resistant. Fiberglass reinforced plastic (FRP) tanks are routinely used for storage of concentrated HCl, but require special construction employing a thicker chemically resistant veil. Otherwise, sudden and unexpected failures may occur. An occasional failure has been caused by chlorinated solvent contamination in the acid.

Inorganic. Glass and other ceramic materials will withstand HCl up to the atmospheric boiling point, above which some attack is to be expected.

Pitfalls

The most common pitfall in handling HCl is the unrecognized presence of oxidizing contaminants, which adversely affect corrosion resistance of both nickel-molybdenum alloys and of zirconium.

Traces of organic solvents may ruin the resistance of both rubber-lined and FRP equipment.

Glass-lined equipment will fail due to *external* corrosion from HCl spillage or vapors, because the corrosion of steel generates hydrogen which penetrates from the outside in the atomic state and then dimerizes to molecular hydrogen at the steel-glass interface. The resulting pressure will pop off the glass lining locally. (*Note:* Glass-lined vessel *jackets* should not be cleaned with inhibited HCl for the same reason.)

Storage and Handling

Following is a listing of materials considered to be adequate for the service intended:

Tanks	FRP or rubber-lined steel
Tank trucks	Rubber-lined steel
Railroad cars	Rubber-lined steel
Piping	FRP or plastic-lined steel (e.g., PP)
Valves	PTFE- or FEP-lined; cast alloy B* (N10001; N-12M)
Pumps	PTFE-lined; impervious graphite; cast alloy B*
Gaskets	Rubber, felted PTFE, FEP envelope, flexible graphite

References

1. T. F. Degnan, "Hydrochloric Acid and Hydrogen Chloride," *Process Industries Corrosion—The Theory and Practice*, NACE, Houston, 1986.

---

*Requires acid free of iron or other oxidizing contaminants

12

Hydrocyanic Acid

Introduction

Hydrogen cyanide is a colorless, poisonous, low-viscosity liquid, with the odor of bitter almonds. The aqueous solution of hydrogen cyanide (HCN) is known as hydrocyanic (or prussic) acid. Hydrogen cyanide is a very poisonous gas (HCN generated by the action of acid on sodium cyanide is used for execution in the gas chambers of some states), and the aqueous solutions and salts are also poisonous if ingested or absorbed. Extreme precautions must be taken in handling HCN and its derivatives.

Water solutions of HCN cause SCC of carbon steel. Corrosivity of hydrogen cyanide relative to steel and stainless steels is complicated by the fact that sulfur dioxide and sulfuric acid (at 50 ppm) are added to the vapor and liquid phases, respectively, to control polymerization. Solutions containing sulfuric acid corrode steel above about 40°C (105°F) and stainless steels above about 80°C (175°F). The care with which such inhibition is effected relates to some discrepancies in corrosion rates reported in the literature. On the other hand, HCN *is* corrosive not only to copper but even toward such noble metals as silver and gold.

Process

The older processes (e.g., release from sodium cyanide, dehydration of formamide) are no longer used, because of the high cost of raw materials. There are basically two types of processes extant;

1. Methane-ammonia-air (the Andrussow process)

2. Methane-ammonia (the BMA process)

HCN is also a by-product of the acrylonitrile process, and small amounts are also recovered from coke oven gases.

Methane-ammonia-air process (Fig. 12.1)

Carried out at about 1100°C (2000°F) over a precious metal catalyst, this process produces HCN by direct oxidation in stainless-steel converters.[1] The reaction mixture is quickly quenched to less than 400°C (750°F), to minimize decomposition of product, with a waste-heat boiler. Unreacted ammonia is preferably recycled via a monoammonium phosphate absorption system. (Ammonia may also be absorbed in sulfuric acid, but disposal of the ammonium sulfate

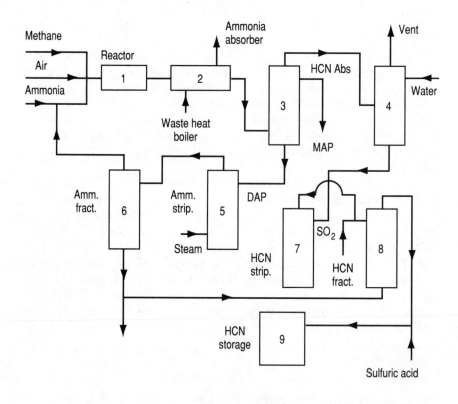

1. S31003   6. S31603
2. S31603   7. Aluminum (A95154)
3. S31603   8. A95154
4. S31603   9. Carbon steel
5. S31603

Figure 12.1   Hydrogen cyanide.

may constitute an environmental problem.) Wastewater is treated by alkaline chlorination to convert cyanide to cyanates.

The HCN product is absorbed in sulfuric-acid-acidified water, stripped from the water, and inhibited with both sulfur dioxide and sulfuric acid for storage. The Andrussow process is the predominant method of production.

Methane-ammonia

In this process, the gas-ammonia mixture is pyrolized in sintered alumina tubes lined with 70% platinum at 1200 to 1300°C (2200 to 2370°F). The reaction is

$$CH_4 + NH_3 = HCN + 3H_2$$

This process is suitable where a use for the by-product hydrogen exists.

Materials

Aluminum alloys

Aluminum is one of the most resistant materials and is used in towers, condensers, heat exchangers, tanks, drums, and piping. The specific alloy depends on strength requirements and available forms (e.g., A91100, A93003, A95154, etc.).[2]

Steel and cast iron

Steel is resistant at ambient temperatures (contingent on proper control of the inhibition). Cast iron is resistant likewise, but is not recommended because of the danger inherent in its brittle nature.

Stainless steels

High-temperature grades (e.g., S31050) are used in converters, and molybdenum-bearing grades (S31603) are used in phosphate solutions.[3] In HCN, rates are less than 1 mpy at 7°C (45°F). However, the stainless steels are no better than steel for storage purposes, showing both less than 2 mpy at ambient temperatures and less than 20 mpy in the 38 to 99°C (100 to 200°F) temperature range, probably because of the effects of stabilizing agents.

High-performance grades (e.g., N08020) have been used to resist sulfuric acid effects under process conditions.

Copper and its alloys

Copper and its alloys resist 100% HCN to at least 200°C (390°F) but are severely corroded by aqueous solutions.

Nickel alloys

The nickel-copper alloy N04400 has been used for drums, piping, etc. for pure HCN, but is also attacked by aqueous solutions. The higher nickel alloys are not more resistant and are not reported in use in this service. Both N02200 and N06600 show less than 20 mpy in the 10 to 38°C (50 to 100°F) range.

Reactive metals

Titanium is resistant to about 100°C (212°F), but is not reported to be used in this service. Titanium and tantalum apparently resist cyanide salts; data on zirconium are lacking.

Precious metals

HCN and cyanide solutions are corrosive to gold in the presence of oxygen or oxidizing agents.

Other metals and alloys

Lead has been used to contain HCN, but this was in the sodium cyanide–sulfuric acid process, and lead alloys are not used in modern production.

Nonmetallic materials

Organic. Some plastic materials are suitable for handling HCN, e.g., PVC and PE to 60°C (140°F), PVDF to 135°C (275°F).[4] Because of toxicity considerations, these should be used only as plastic-lined pipe.

Of the elastomers, only butyl rubber would be considered, with a temperature limitation of 66°C (150°F).[4]

Inorganic. Glass or glass-lined equipment is useful to about 100°C (212°F).

Pitfalls

SCC of steel may occur if vessels are not properly stress-relieved after welding or forming.

Storage and Handling

HCN is stored and shipped as 99.5% concentration minimum (0.5% water max.; 0.06 to 0.10% acidity). Sulfuric acid (or phosphoric) and sulfur dioxide are added as stabilizers to prevent polymerization.

The following materials of construction are common for handling and storage, and their use is thought to constitute good engineering practice:

Tanks	Carbon steel cylinders (4.5, 22.7, 45.4 kg)
Tank trucks	Steel, aluminum, or stainless steel
Railroad cars	Steel, aluminum, or stainless steel
Piping	Aluminum or N04400
Valves	CF3M
Pumps	CN7M
Gaskets	Butyl rubber, spiral-wound stainless steel/PTFE

References

1. P. J. Carlisle, "Manufacturing, Handling and Use of Hydrocyanic Acid," *Trans. AIChE*, vol. 29, 1933.
2. J. A. Lee, "Materials of Construction for the Chemical Process Industries," McGraw-Hill Book, New York, 1950.
3. J. A. Lee, "Hydrogen Cyanide Production," *Chemical Engineering*, vol. 56, no. 2, 1949, pp. 134–136.
4. I. Mellan, *Corrosion-Resistant Materials Handbook*, Noyes Data Corp., Park Ridge, N.J., 1976.

Hydrofluoric Acid

Introduction

Hydrofluoric acid, a relatively weak reducing acid, is a water solution of hydrogen fluoride. Commercially, it is used for pickling operations, acid treatment of oil wells, and etching glass. The anhydrous hydrogen fluoride (AHF) is used in chemical processes (e.g., manufacture of fluorinated organic compounds, alkylation of petroleum products).

Both forms are *extremely* hazardous, inflicting painful burns. Breathing the vapors causes lung damage, and the Occupational Safety and Health Administration (OSHA) limits exposure to a time-weighted average of not more than 3 ppm during a 40-hour workweek. Fluoride salts are toxic in *high* concentrations, although fluoridation of water supplies is a standard and safe practice for dental hygiene.

The acid, although monobasic, can link molecules in a chainlike configuration (H-F-H-F-, etc.), permitting the formation of acid salts (e.g., ammonium bifluoride, NH_4HF_2).

Process

Fluorspar, a crude calcium fluoride with siliceous contaminants, and 98 to 99% sulfuric acid generate 70 to 75% HF in a steel kiln, often lined with N10276 or N06030, at about 150°C (300°F), discharging calcium sulfate to waste (Fig. 13.1). The offgas is preconditioned, scrubbed with sulfuric acid and stored as 80% acid. Vent gases are further scrubbed with water, the tails being recovered as a 30 to 35% fluosilicic acid. Oleum is added as makeup acid to the reactor.

Except for the kiln and some high-nickel alloy internals, carbon steel can be used up to the crude acid condenser, which is of alloy C276 (N10276) construction. The acid scrubber is of impervious graphite, as is the water scrubber.

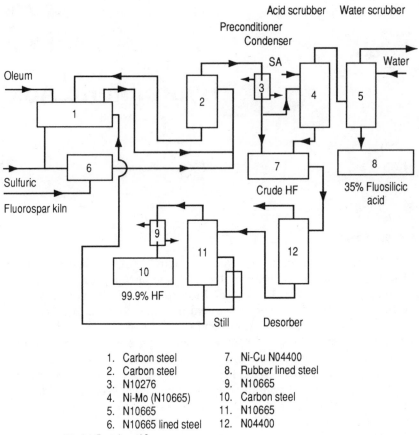

Figure 13.1   Hydrofluoric acid.

1. Carbon steel
2. Carbon steel
3. N10276
4. Ni-Mo (N10665)
5. N10665
6. N10665 lined steel
7. Ni-Cu N04400
8. Rubber lined steel
9. N10665
10. Carbon steel
11. N10665
12. N04400

The 80% HF is desorbed of sulfurous oxides and distilled to produce anhydrous HF, in alloy 400 or copper columns and alloy 400 refrigerated condensers for storage. Commercial storage is in steel tanks.

Materials

Essential information for the safe handling and use of HF is contained in the Manufacturing Chemists Association's "Chemical Safety Data Sheet SD-25." Up-to-date information is available in the manufacturers' latest product bulletins.

Aluminum alloys

Aluminum and its alloys should not be exposed to any concentration of hydrofluoric acid.

Cast iron and steel

Cast iron should not be used in HF service both because of the hazards associated with its brittle properties and because of selective attack on siliceous components. Even the 14% silicon cast irons are nonresistant, because of the attack on siliceous films.

Carbon steel is conventionally used with HF above 64%. The nearest commercial grade is 70% HF, and the recommended maximum temperature is 32°C (90°F).[1]

Stainless steels

The martensitic and ferritic grades are considered unsuitable because of hydrogen-assisted cracking (HAC) and a sensitivity to velocity effects.[2]

Austenitic stainless steels are unreliable in dilute HF, being subject to SCC under some conditions, but type 304 resists AHF up to at least 100°C (212°F). The equivalent cast alloy CF8 (UNS J92900) has been used for pumps.

The nickel-rich high-performance grades, such as CN7M (J95150 or N08007), resist 70% acid well, and the wrought alloy (N08020) is used as valve trim in AHF. Alloys N08825 and N06985 may be used in lieu of N04400 (see below) to avoid SCC.

Copper alloys

Copper is satisfactory, in the absence of oxygen and/or sulfur dioxide contamination, being a close second to alloy 400 (see below) for HF service. It is especially useful for flexible tubing connections and as a distillation column for AHF. Copper is, however, also sensitive to velocity effects.[2]

Nickel alloys

Alloy 200 (N02200) is resistant to aqueous HF, but less so than alloy 400. It is limited to air-free systems below 80°C (175°F), and is not economically attractive.

Alloy 400 (N04400) is the preferred nickel-base alloy for all concentrations of HF up to about 120°C (250°F). Both the 400 and 500 alloys are susceptible to SCC in moist vapors in the presence of traces of oxygen. Thermal stress-relief does not prevent SCC by traces of cupric fluoride.

Alloy 600 (N06600) is sometimes employed in valves for AHF to avoid the potential SCC problem with alloy 400.

The nickel-molybdenum alloy B-2 (N10665) is resistant to sulfuric-HF mixtures encountered in the manufacturing process.

Reactive metals

The reactive metals (i.e., titanium, zirconium, and tantalum) are non-resistant to even traces of HF.

Noble metals

Silver, gold, and platinum are all suitably resistant to HF and have been used in specialty equipment as needed.

Other metals

Lead is attacked by hydrofluoric acid solutions.

Nonmetals

Uses and limitations of plastics, elastomers, and carbon-graphite have been detailed.[2]

Organic. Plastic materials *without hydroxyl groups* tend to resist HF within certain parameters (HF is strongly hygroscopic, rapidly attacking polyesters, for example).

PVDF, PE, and PP are resistant to 70% acid to about 32°C (90°F). The higher fluoropolymers [PTFE, PFA, FEP, chlortrifluorethylene (CTFE)] are resistant to their normal temperature limits, but are subject to permeation with attendant hazard to adjacent metals.

Synthetic soft rubbers (butyl, neoprene) will withstand 60% HF to about 70°C (160°F). (Compounding is critical; silica and magnesia must not be used.)

Inorganic. Of the inorganic materials, glass and siliceous ceramics are not resistant.

Carbon and graphite are resistant, but the impregnants may be attacked. Impervious graphite is limited to 48% acid at the boiling point and to 85°C (185°F) in 60% HF.

Pitfalls

High-strength steels are subject to hydrogen embrittlement and HAC by aqueous HF. High-strength steel bolting (e.g., B-7 bolts) has failed in flanges by environmental cracking, because of leakage of HF.

Nominally resistant stainless castings (e.g., CF8) have corroded because of direct impingement of AHF. Traces of chloride contamination can cause SCC of austenitic stainless steels.

Alloy 400 is susceptible to SCC in HF vapors in the presence of air or oxidants, apparently due to formation of a film rich in cupric fluoride.

Storage and Handling

Materials to be considered for handling and storage of HF in concentrations greater than 65% at ambient temperatures are as follows:

Tanks	Steel
Tank trucks	Steel
Railroad cars	Steel, butyl- or chlorobutyl-rubber-lined (to 32°C maximum)
Piping	Alloy 400 or PTFE-lined steel
Valves	PTFE-lined, alloy 400, steel with N04400 or N0802 trim
Pumps	Alloy 400
Gaskets	Flexible graphite; spiral-wound alloy 400/PTFE

References

1. T. F. Degnan, "Materials of Construction for Hydrofluoric Acid and Hydrogen Fluoride," *Process Industries Corrosion—The Theory and Practice*, NACE, Houston, 1986.
2. "Materials for Receiving, Handling and Storing Hydrofluoric Acid," 1983 Revision, item No. 54201, Publication 5A171, NACE, Houston.

Chapter

14

Hydrogen Peroxide

Introduction

Hydrogen peroxide is a powerful oxidizing agent, a characteristic that affects both corrosion and compatibility with (i.e., product stability in) materials of construction. Stability of H_2O_2 increases with concentration.

Pure hydrogen peroxide is inherently stable when purified and stored in clean, nonreactive containers. Decomposition is caused catalytically by trace amounts (e.g., parts per *billion*) of such metals as lead, manganese, copper, and iron. Fortunately, since it is almost impossible to totally prevent such contamination, this effect can be offset by certain inhibitors which stabilize the catalytic contaminants (see "Materials" below).

Process (Fig. 14.1)

Traditional processes include acidification of inorganic peroxides, electrolysis of sulfuric acid solutions, and autoxidation of organic compounds.[1]

The predominant modern process is autoxidation of anthraquinone. The complex aromatic compound is catalytically reduced (e.g., over palladium) at about 50°C (120°F) and subsequently oxidized to release derivative compounds and hydrogen peroxide. The peroxide is extracted with water and purified by distillation under reduced pressure, the organic stream being recycled. The chemistry and process control requirement are very complex (see *Encyclopedia of Chemical Technology,* John Wiley & Sons, New York).

In an older process, potassium bisulfate, together with potassium sulfate and sulfuric acid, undergoes electrolysis to form potassium persulfate as an intermediate product. The solid persulfate is then hy-

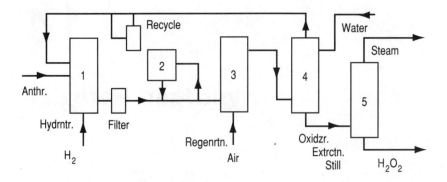

1. S30403
2. S30403
3. S30403
4. S30403
5. Aluminum A91100

Figure 14.1   Hydrogen peroxide by autoxidation.

drolyzed with steam and sulfuric acid, a 35% H_2O_2 product (Albone) being distilled in a stoneware column using aluminum heating coils and reflux condenser. The 35% peroxide is subsequently vaporized in stoneware retorts with a type 317L (S31703) coil, concentrated by distillation in chemical stoneware, condensed in S31703 exchangers, and stored in aluminum tanks (Fig. 14.2).[2]

Materials

Aluminum alloys

Aluminum is resistant to hydrogen peroxide at the neutral point, although general corrosion is accelerated in both acidic and alkaline solutions.[2] Below the neutral point, corrosion can be inhibited by pyrophosphate, while pitting attack is inhibited by nitrates. For long-term storage, the aluminum should be 99.6% pure (A91100) and free of surface contamination by foreign matter of any kind. A preservice pickling with about 40% nitric acid is recommended. Aluminum storage tanks improve with age, apparently due to progressive reinforcement of the protective aluminum oxide film. For short-term storage (e.g., pipe, fittings, tank cars), grades of lower purity (i.e., A93003 or A95052) are acceptable. Higher-strength alloys may be used for brief contact, *not* storage, as required.

Steel and cast iron

These ferrous metals are usually unacceptable for hydrogen peroxide service. However, properly preconditioned cast-iron equipment finds

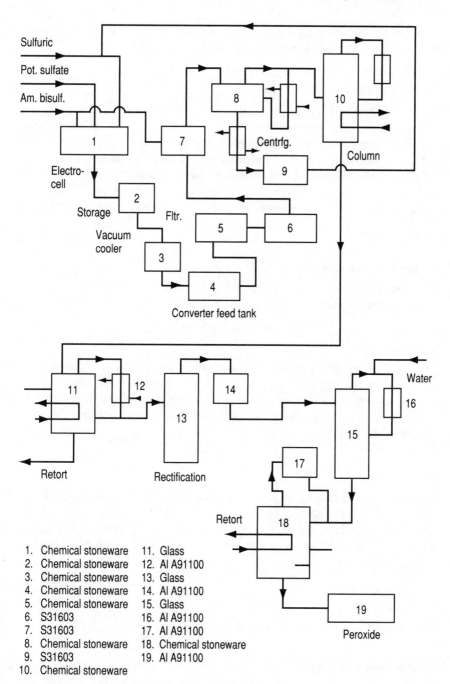

Figure 14.2 Hydrogen peroxide (sulfate electrolysis).

some application in dilute alkaline hydrogen peroxide bleaching operations, with sodium silicate as a corrosion inhibitor.

Stainless steels

Austenitic 18-8 grades of stainless steel, pickled and passivated with warm dilute nitric acid, are satisfactory for handling solutions of less than 8% concentration. Stainless-steel tanks and agitators of type 304L (S30403) are used to handle dilute peroxide bleaching solutions.

Because of their ease of fabrication and procurement, they are convenient to use for stronger solutions, but for short-time handling only, as in valves, pumps, and pipelines, despite a mildly adverse catalytic effect.

Copper and its alloys

Cuprous alloys are not suitable for peroxide service, since they undergo corrosion and decompose the product.

Nickel alloys

Alloy 400 (N04400) has found some useful applications in dilute peroxide bleaching operations, despite its copper content. In general, chromium-free nickel-base alloys are unsuitable in peroxide service. Chromium-bearing grades (e.g., N10276) are highly resistant but not economical.

Reactive metals

Titanium is resistant, as is zirconium, to at least 50% peroxide at 100°C (212°F). Tantalum is very resistant to peroxide over a wide range of conditions relative to both corrosion and product stability. Such materials are rarely required and cost more than competitive materials.

Precious metals

Precious metals find no application in this service.

Other metals and alloys

Zinc is corroded by peroxide. Lead is attacked, with pitting in 50% solutions. Such metals also accelerate decomposition of the peroxide.

Nonmetallic materials

Organic. Such solid plastics as PVC, CPVC, PE, chlorinated PE, vinylidene fluoride/hexafluoropropylene, and various fluorinated plastics have been used for various concentrations of hydrogen peroxide.

Rubber and elastomers, unless specially compounded, are deteriorated by the peroxide, which is simultaneously decomposed. Buna-N is the most resistant elastomer, except for vinylidene fluoride/hexafluoropropylene. In dilute alkaline solutions, specially compounded natural rubber or neoprene hoses may give several years' life.

Organic coatings are generally unsuitable, because of corrosion of the substrate metal at "holidays" in the coating.

Wooden tanks and piping, made of cypress and pine, have been fairly satisfactory for dilute solutions under many conditions. At high temperatures and alkaline conditions, wood is swelled, softened, and delignified. For satisfactory use, wooden tanks should have a tile or concrete bottom (see below).

Inorganic. Glass, glass-lined steel, and ceramicware are excellent for hydrogen peroxide service, when properly cleaned. Acidproof tile and tile-lined concrete tanks, assembled with acid- and alkali-resistant cement, are useful in bleaching operations.

Concrete is not recommended for strong peroxide but has been useful in dilute alkaline bleaching solutions, particularly when sodium silicate inhibitors are present.

Pitfalls

The primary pitfall in handling hydrogen peroxide is unanticipated surface contamination. Oil, grease, metal oxides, etc. accelerate decomposition. Cleanliness must be scrupulously observed. Welds should be properly effected by using inert-gas-shielded welding to prevent degradation by welding slag or flux residues.

Steel contamination *must* be avoided. A rusted steel pipe plug in a stainless line beneath a storage tank caused a tank to overflow with frothy peroxide when the bottom valve was opened.

Storage and Handling

Commercial hydrogen peroxide is usually marketed as a 50 to 70% water solution. The Interstate Commerce Commission (ICC) classifies solutions over 8% by weight as corrosive. The CP reagent-grade product is 35%. Solutions over 52% are shipped in *double-headed* aluminum drums or aluminum trucks or tank cars. The 70% solutions are stabilized to cope with dilution to 35% or 50% concentration.

The following materials of construction are commonly used for handling and storage, and their use is thought to constitute good engineering practice:

Tanks	Aluminum A91100, A93003, or A95052
Tank trucks	Aluminum A91100 or A93003
Railroad cars	Aluminum 1100 or 3003
Piping	Aluminum 3003 or stainless S30403
Valves	Aluminum, CF3M, or CN7M
Pumps	Aluminum or CN7M (with vinylidene fluoride/hexafluoropropylene seals)
Gaskets	Aluminum, spiral-wound PTFE/stainless, flexible graphite, buna N.

References

1. J. E. Lee, *Materials of Construction for Chemical Process Industries*, McGraw-Hill, New York, 1950.
2. J. S. Reichert and R. H. Pete, *Chemical Engineering*, vol. 54, no. 4, 1947, pp. 213–228.

15

Hydrogen Sulfide

Introduction

Hydrogen sulfide is a colorless, highly poisonous gas having the characteristic odor of rotten eggs. It is used in certain chemical processes, e.g., extraction of deuterium oxide ("heavy water") from natural water. It is also commonly recovered from sulfidic ("sour") petroleum products. It can be a by-product of other processes [e.g., fuel cells, coal conversion, synthetic natural gas (SNG), natural gas, pulp mills, geothermal steam]. Hydrogen sulfide is rarely used as a chemical compound but is an important precursor in some sulfuric acid and sulfur production processes.

Process

One common source of recovered hydrogen sulfide is an acid-gas scrubbing system for sour gas, e.g., a 25% monoethanolamine (MEA) system (Figure 15.1).

The gas stream passes to a pressurized absorber of stress-relieved carbon steel construction in which the hydrogen sulfide (together with any coexistent carbon dioxide) is reacted with a water solution of MEA. The rich solution enters an austenitic stainless steel or impervious graphite interchanger which preheats the feed while cooling the stripping still tails. The stripping still overhead is condensed with water reflux, the uncondensed acid gases being passed to a compressor for further utilization. The tails stream is "lean" 25% MEA, which is further cooled and recycled to the absorber.

Other amines are also used in similar systems, e.g., diethanolamine (DEA), diglycolamine (DGA).

Materials

Because of hydrogen-assisted cracking and other phenomena related to nascent atomic hydrogen generated by corrosion in H_2S systems, mate-

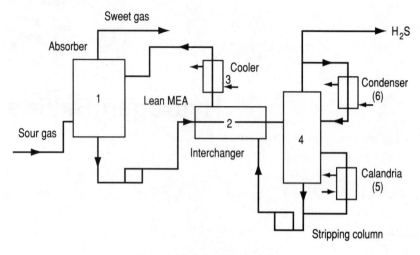

Figure 15.1 Hydrogen sulfide from MEA system.

rials selection and quality control (e.g., stress-relief, hardness limitations) are critical. Requirements are discussed in Standard MR0175.[1]

Aluminum alloys

Aluminum alloys are fully resistant to wet or dry hydrogen sulfide gases and to sulfidic waters. Aluminum coatings (i.e., Alonizing or Calorizing) are used to protect steel from sulfide attack at elevated temperatures.

Steel and cast iron

Steel is reasonably resistant to general corrosion by wet hydrogen sulfide, but the protective iron sulfide film is easily removed by erosion or impingement. Note that iron sulfides may be pyrophoric, a potential source of fire or explosion on exposure to air (see Chap. 66). Hydrogen blistering may occur in wet sulfide service, due to manganese sulfide inclusions within the steel.

Hardened steel (above about 22 Rockwell C hardness [HRC]) is susceptible to sulfide stress cracking (SSC; see Chap. 60).

Gray cast iron will suffer graphitic corrosion in acidic hydrogen sulfide streams, but austenitic nickel cast irons (e.g., F43000) are successfully used for pumps handling sulfidic waters. Rates for the nickel cast iron are 15 to 20% of the rates for gray cast iron in wet hydrogen sulfide.

Stainless steels

The austenitic stainless steels are corrosion-resistant to both wet and dry hydrogen sulfide. (See "Pitfalls" below.)

Copper and its alloys

High copper alloys react with wet hydrogen sulfide, forming voluminous black corrosion-product films. Yellow brasses form more adherent and protective films.

Nickel alloys

At temperatures below about 260°C (500°F), alloys 200 (N02200) and 400 (N04400) may be used as required. Chromium- and molybdenum-bearing nickel-rich and nickel-base alloys find application as in oil-drilling operations "down-hole," a severe sulfide/chloride environment. However, nickel and its alloys are not used for conventional handling of hydrogen sulfide because other, less expensive materials are suitable.

Reactive and refractory metals

Titanium has been used for severe sulfidic services involving chloride/sulfide corrosion. However, zirconium and tantalum, although resistant, find no significant applications.

Precious metals

Silver is subject to tarnishing in wet hydrogen sulfide exposures. Gold and platinum are not used in such applications.

Other metals and alloys

Zinc reacts with hydrogen sulfide, and sulfide surface contamination is a problem with zinc-pigmented primers.

Lead suffers light-to-moderate attack (less than 20 mpy) in wet hydrogen sulfide.

Nonmetallic materials

Organic. Impervious graphite heat exchangers have been used to handle saturated sulfidic waters up to the atmospheric boiling point.

A large number of thermoplastic and thermosetting resinous materials (PE, PP, PVC, CPVC, FEP, epoxies, phenols, furans, polyesters) resist hydrogen sulfide and sulfidic waters, within their normal temperature limitations.

Rubbers and elastomers are fairly resistant. Neoprene and chlorinated rubber are satisfactory at ambient temperature, and butyl rubber to about 66°C (150°F). Elastomers undergo further uncontrolled vulcanization at higher temperatures.

Inorganic. Glass and other ceramics resist wet hydrogen sulfide but conventional cement or concrete undergoes acid attack.

Pitfalls

Hydrogen sulfide is *extremely* toxic (even more so than hydrogen cyanide), 3000 to 4000 ppm in air causing instant death. It is the more dangerous in that it paralyzes the olfactory nerves so that the victim is less conscious of the characteristic rotten-egg odor.

Pyrophoric iron sulfide may be formed on iron or steel exposed to wet hydrogen sulfide, while hydrogen embrittlement (see Chap. 52), blistering, or SSC may occur under some conditions. Slow-strain-rate embrittlement is a particular problem in storage of wet hydrogen sulfide in steel tanks or vessels.

The 18-8 stainless steels are subject to SCC by chloride-contaminated sulfidic waters. A saturated 500-ppm chloride water will cause SCC of highly stressed S30403 at room temperature within 24 hours. In the *sensitized* condition, S30400 suffers intergranular stress-corrosion cracking (IGSCC) when exposed to sulfidic waters contaminated with air (which forms polythionic acids).

Storage and Handling

The following materials of construction are commonly used for handling and storage, and their use is thought to constitute good engineering practice for dry H_2S:

Tanks	Steel
Tank trucks	Steel
Railroad cars	Steel
Piping	Steel
Valves	CF8M
Pumps	CF8M
Gaskets	Neoprene, spiral-wound PTFE/S30400

References

1. Standard MR0175, "Sulfide Stress Cracking-Resistant Metallic Materials for Oilfield Equipment," NACE, Houston, 1990.

Magnesium Chloride

Introduction

The hexahydrate salt of magnesium chloride ($MgCl_2 . 6H_2O$) has little commercial use, but the salt is the major source material in the production of magnesium metal. Magnesium chloride is a by-product in the production of hafnium by magnesium reduction of hafnium tetrachloride.

Process

Magnesium chloride is produced both from salt brines and from seawater, in which it amounts to about one-eighth of the salt content. It is also manufactured in the anhydrous form by direct chlorination of magnesia in the presence of powdered coke or coal (which complete the reduction of oxychlorides produced as by-products).[1]

Brine process

Natural brines are first treated to remove bromine, then traces of iron are removed, and the solution is evaporated to precipitate a maximum amount of sodium chloride. The salt is physically removed and the mother liquor evaporated under controlled conditions, precipitating the double salt $2MgCl \cdot CaCl_2 \cdot 12H_2O$ (*tachydrite*). The tachydrite is richer in magnesium than calcium, depleting liquor of magnesium until a ratio of 1 Mg to 10 Ca is attained.

The precipitated tachydrite is dissolved in hot water and subsequently cooled to produce substantially pure crystalline magnesium chloride hexahydrate.

Materials of construction for crystallizers are brick-lined above about 80°C (175°F) or rubber-lined below. Dryers, evaporators, thickeners, and tanks are brick-lined, while associated hardware, piping, and heat exchangers are alloy 600 (N06600) or 625 (N06625). Pumps are nickel cast iron (F43000). In some plants, glass-lined equipment is preferred to resist the acidic salt and traces of hydrochloric acid released by hydrolysis.

Seawater process

Seawater is treated with lime in a concrete tank, precipitating magnesium hydroxide which is passed to steel thickeners and filtered. The filter cake is transferred to a rubber-lined tank and reacted with 10% HCl to form magnesium chloride. The dilute magnesium chloride solution is sprayed into a brick-lined furnace from which steam, water, and combustion gases pass overhead while a 35% magnesium chloride solution is removed as tails to brick-lined retention tanks containing high-alloy stainless heating coils (e.g., N08825, N08020) to resist SCC.

Finally, a 48% solution is dehydrated in a shelf drier to solid granules as the final product.

Materials

Aluminum alloys

Because of the acidity and iron contamination, aluminum finds no application in this process. Dilute solutions (1 to 10%) cause pitting.

Steel and cast iron

Steel and iron can be used when iron contamination is not objectionable and acidity is low, but acid attack is a constant threat.

Austenitic nickel cast iron (F43000), trimmed with alloy 600, has been used for pumps and valves in some applications.

Stainless steels

Conventional type 316L, stress-relieved, was used in early years, but has been largely supplanted by SCC-resistant high-performance alloys (S31254, N08904, N08367).

Copper and its alloys

Copper alloys are not suitable for this service.

Nickel alloys

Chromium-bearing nickel-base alloys have been standard in this service in lieu of stainless steels, because of much greater resistance to

SCC. The molybdenum-bearing grades (e.g., N06625, N06022, N10276) are the most reliable against chloride pitting attack as well.

Reactive metals

Titanium has no useful applications in this service, because of high acidity and low oxidizing capacity, although laboratory tests show it resistant to the standard boiling 42% $MgCl_2$ test. Zirconium and tantalum could be employed where needed.

Precious metals

Silver, gold, and platinum find no application in this service.

Other metals and alloys

Lead is attacked at 20 to 50 mpy by magnesium chloride solutions, but has been used in dilute (less than 10%) solutions.

Nonmetallic materials

Organic. Various types of plastic are useful against acid chloride salt solutions, within the permissible temperature and pressure limitations. Rubbers and elastomers are unaffected, with Buna N and neoprene having the higher temperature limitation of 93°C (200°F).[2]

Impervious graphite heat exchangers are suitable for heating the salt solutions.

Asphalt-impregnated wood has been used to handle acid-contaminated steam and water from the combustion products of the magnesium chloride furnace.

Inorganic. Glass, ceramicware, and acid-proof brick and cement are all useful in magnesium chloride equipment, resisting HCl attack.

Ordinary concrete is severely attacked by weak acids as well as by the specific in-situ replacement of calcium by magnesium (corrosion II).

Pitfalls

The major pitfalls are associated with uncontrolled acidity, which causes corrosion of iron and steel, and the dangers of chloride pitting and SCC in susceptible materials.

Storage and Handling

The following materials of construction are commonly used for handling and storage, and their use is thought to constitute good engineering practice:

Tanks	Rubber-lined steel
Tank trucks	Rubber-lined steel
Railroad cars	Rubber-lined steel
Piping	Alloy 600
Valves	Nickel cast iron or CN7M
Pumps	Nickel cast iron or CN7M
Gaskets	Elastomers or felted PTFE

Solid product is handled in steel, when a small amount of corrosion and contamination is acceptable, or in polyethylene drums with a steel overpack.

References

1. J. A. Lee, *Materials of Construction for Chemical Process Industries*, McGraw-Hill, New York, 1950.
2. I. Mellan, *Corrosion Resistant Materials Handbook*, Noyes Data Corp., Park Ridge, N.J., 1976.

17

Magnesium Sulfate

Introduction

Magnesium sulfate (*Epsom salts*) is used in dyeing and printing, in processes for chrome tanning, as a supplement in fertilizers, and in other chemical and pharmaceutical applications. It is handled and packaged as the solid heptahydrate crystal $MgSO_4 \cdot 7H_2O$. It is found in the natural state, being mined as the mineral langbeinite, as well as being a manufactured product.

As the salt of a strong acid and a weak base, magnesium sulfate in aqueous solution acts like dilute sulfuric acid, unless pH-adjusted or influenced by specific oxidizing or reducing contaminants.

Process

The product cannot be directly crystallized, but must be made by dehydration of a hydrate. Aqueous solutions are prepared by reaction of magnesium oxide, hydroxide, or carbonate in sulfuric or sulfurous acid.

One process involves lime treatment of a high-purity magnesium chloride solution (see Chap. 16, "Magnesium Chloride") to precipitate magnesium hydroxide, which is then treated with sulfuric acid. Iron, aluminum, and other impurities are precipitated by reaction with excess magnesium oxide. The resulting solution is vacuum-crystallized to yield the United States Pharmacopeia (USP) product as commercial Epsom salts.

Materials

Aluminum alloys

Aluminum is satisfactory for handling Epsom salts in either the dry state or solution. Air-saturated 50% solutions at 60° to 70°C (140° to

160°F) in silk processing are handled in aluminum to avoid product discoloration.[1]

Aluminum is not resistant to crude magnesium sulfate solutions, because of their acidic nature combined with iron contamination.

Steel and cast iron

Ordinary steel has been used for agitators, bins and hoppers, centrifuges, heat exchangers, conveyors, crystallizers, driers, filter presses, piping, settling and storage tanks, and thickeners.[2] It is also satisfactory for Epsom salts, dry or in solution.

Cast iron has been used to handle the product.

Stainless steels

Various grades of stainless steels, but especially type 316L (S31603) and molybdenum-bearing high-performance grades (alloy 20Cb3 or N08020), are used to resist the acidic conditions analogous to weak sulfuric acid.

Copper and its alloys

Copper alloys are unsuitable because of corrosion by aerated acidic solutions and color contamination. Corrosion is autocatalytic because of accretion of cupric ions.

Nickel alloys

Alloy 400 (N04400) has been successfully used for evaporators, heating coils and tubes, crystallizers, extractor baskets, and driers.

Reactive metals

The reactive metals find no application in this service, although titanium is resistant at room temperature and zirconium and tantalum at higher temperatures.

Precious metals

There is no application for precious metals in this service.

Other metals and alloys

Lead is resistant (less than 20 mpy) in 10 to 60% solutions to about 100°C (212°F).

Nonmetallic materials

Organic. Wooden tanks, piping, etc. are satisfactory for the weakly acidic magnesium sulfate solutions.

Conventional plastics and elastomers are resistant up to their normal temperature and pressure limitations.

Impervious graphite exchangers are suitable for heating magnesium sulfate solutions.

Inorganic. Glass and ceramicware are resistant, but concrete is severely attacked by magnesium sulfate. Corrosion II occurs by reaction of magnesium with the cement, leaving a gelatinous residue of a nonbinding nature, while the sulfate ion causes sulfate bacillus (corrosion III) attack.[3]

Pitfalls

Magnesium sulfate solutions pose a danger of SCC of the austenitic stainless steels in the event of chloride contamination.

Storage and Handling

Epsom salts are packaged in wooden barrels and kegs, moistureproof bags, and glass or plastic containers.

References

1. I. Mellan, *Corrosion Resistant Materials Handbook*, Noyes Data Corp., Park Ridge, N.J., 1976.
2. J. A. Lee, *Materials of Construction for Chemical Process Industries*, McGraw-Hill, New York, 1950.
3. C. P. Dillon, *Corrosion Control in the Chemical Process Industries*, McGraw-Hill, New York, 1986.

Nitric Acid

Introduction

Nitric acid is a strong mineral acid, highly oxidizing in nature. The conventional ammonia oxidation process yields a product of about 60% concentration, which can be purified and concentrated to give CP acid of 70% concentration. Very strong acid (i.e., 90 to 100%) is made by dehydrating the more dilute product with concentrated sulfuric acid. (Acid above 85% strength is called *fuming nitric acid* because of the vapors of red or white oxides of nitrogen.)

Process (See Fig. 18.1)

Anhydrous ammonia is mixed with preheated air under about 690 kPa (100 lb/in^2) pressure and passed over a platinum-rhodium catalyst at about 925°C (1700°F). The reaction product is cooled (preheating the air to the converter in an interchanger, followed by water cooling), and the oxides of nitrogen are absorbed in demineralized water to form 60 to 65% nitric acid. Acid content of 65 to 67% is the azeotrope with water at atmospheric pressure.[1] In the pressure oxidation process, a turbine gas heater is used in connection with the gas turbine driving the air compressor.[2]

Further concentration is effected by addition of concentrated sulfuric acid in a mixing tank (Fig. 18.2). The mixed acids feed a dehydrating tower, 99% acid being condensed from the overhead vapors. The tails are 65 to 70% sulfuric acid containing about 7% nitric acid. The tails stream is stripped of nitric acid in a denitrating column.

Materials

Aluminum alloys

Aluminum alloys are good only for very high acid concentrations, e.g., greater than 80% at room temperature, greater than 93% at 43°C

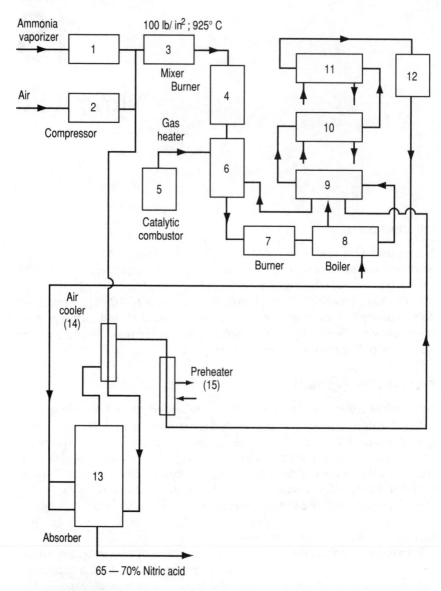

Figure 18.1 Nitric acid (ammonia oxidation).

Ammonia vaporizer

Air

Compressor

1. Carbon steel
2. Carbon steel
3. Ni-Cr (N06600)
4. N06600
5. S30403
6. S30403
7. S30403
8. – 15. S30403

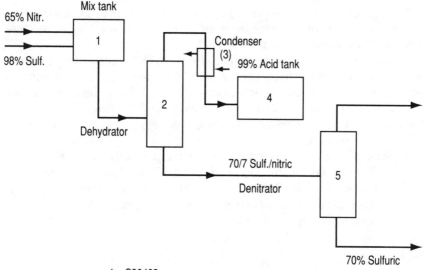

1. S30403
2. 25 Cr–20 Ni–6 Mo (S31254)
3. Ni–Cr–Fe–Mo (N08020) or 26-1 (S44625)
4. Aluminim (A93003, A95154, etc.)
5. Acid-brick lined steel

Figure 18.2 Ninety-nine percent nitric acid.

(110°F). Aluminum alloys commonly employed are A91100, A93003, A95052, and A95454. The first two alloys must be welded with A91100 rod, the second two with A95356. (Welds made with high-silicon rods are subject to accelerated preferential attack.)

Cast iron and steel

Historically, iron and steel have been used in some instances in 95% acid or above at room temperature. In modern practice, they are considered unsuitable.

Stainless steels

Conventional low-carbon or stabilized 18-8–type stainless steels (i.e., S30403, S32100, S34700) are the workhorse materials for handling nitric acid over a wide range of temperatures and concentrations up to about 60% at the atmospheric boiling point. Regular carbon grades, e.g., S30400, are subject to intergranular attack when sensitized by welding or by exposure to temperatures in the 425° to 815°C (800° to 1500°F) range. S30403 of improved quality has been developed, containing extremely low C, Si, and P for resistance to weld decay, and known in the United Kingdom as *nitric acid grade* (NAG).[1]

A little-known problem is *vapor-phase* corrosion of 18-8 stainless steels above the liquid level in greater than 90% acid under storage conditions at ambient temperatures.[3]

Above about 90% concentration, wrought 18% chromium stainless steels with 4 to 5% silicon (and about 18% nickel to balance the desired austenitic structure) are superior to S30403 in immersion service. They are used for 95% acid condensers and services above 70% at about 80°C (175°F) or more.[4] Nothing has been published relative to vapor-phase resistance.

At slightly higher temperatures than are tolerable for S30403, type 310L (S31003) has been employed. A NAG variation is available. Nickel-rich alloys like N08800 have also been used in condensers where high-chloride cooling waters are encountered, specifically to resist waterside pitting and SCC.[5]

Copper alloys

Copper and its alloys are totally unsuitable for any kind of nitric acid exposure.

Nickel alloys

Chromium-bearing grades of nickel-rich and nickel-base alloys are used occasionally for specific purposes. (They are not usually economically competitive with the stainless steels.) They are used primarily in the high-heat areas of the oxidation step, typical selections being alloy 601 (N06601), alloy X (N06002), nickel-chrome solid-solution-strengthened (N06004), and alloy 333 (N06333). Basically, all useful alloys in this application contain at least 35% Ni, 20% Cr, and 0.5% Si.[5]

Reactive metals

Titanium (e.g., grade 2, R50400) and Ti-Pd alloy (grade 7, R52400) are resistant below 10% and from 65 to 90%. Unacceptably high rates (i.e., greater than 10 mpy) are observed in boiling acid in the 20 to 50% range. Titanium is not used above 90% because of a danger of SCC, and never in red fuming nitric acid (RFNA). If the water content falls below 1.34% and nitrous oxide exceeds 6%, a dangerous pyrophoric action can occur in red fuming nitric acid (RFNA).[4]

Zirconium is more resistant than titanium up to the atmospheric boiling point for 65% acid (about 125°C or 255°F), and even as high as 230°C (450°F). However, it is subject to SCC above 70% concentration. It is also unsuitable above 90%.[5]

Tantalum is resistant to all concentrations to the atmospheric boiling point, but is not normally economically competitive. It has been used in fuming nitric acid to 5.5 MPa (800 lb/in^2) to 315°C (600°F). For potential applications above the atmospheric boiling point, corrosion testing is recommended.

Noble metals

Gold and platinum are resistant to nitric acid, but silver is rapidly attacked. There are no cost-effective applications for these metals in conventional nitric acid applications.

Other metals and alloys

Lead, zinc, and tin are severely attacked by nitric acid.

Nonmetals

Nonmetallic materials are limited in nitric acid service both by corrosion characteristics and by the powerful oxidizing capacity of the acid.

Organic. PTFE will resist up to 70% acid to about 230°C (450°F), but a limit of about 25°C (75°F) is suggested above that concentration, primarily because the vapors tend to permeate the plastic.

FEP will tolerate up to 100% acid to 200°C (390°F), while PVDF is recommended to about 50% concentration to 52°C (125°F).

Other plastics (e.g., unplasticized PVC, polyesters, polyalkylenes, synthetic rubbers) can be used only in quite dilute acids and at not much over room temperature.

Carbon is a useful material, provided it is free of organic binders, but impervious graphite is limited to 30% acid at 25°C (77°F) and to 10% at 85°C (185°F).

Inorganic. Glass and ceramic materials resist nitric acid up to 65% to at least 125°C (260°F) and, with reduced strength and life, to almost 205°C (400°F). Because of the leaching of iron and siliceous components, the maximum temperature should be limited to 100°C in the 65 to 100% acid range.

Pitfalls

1. Accumulation of hexavalent chromium (Cr^{+6}) ions increases the corrosivity of nitric acid. In the laboratory, this is a problem with long-term (or high corrosion rate) tests. In process exposure, the accumulation occurs in crevices (e.g., tube-to-tubesheet rolled joints) and as end-grain attack, requiring more highly alloyed stainless steels

than conventional S30403 if crevices cannot be avoided. For tank cars and nitration vessels, cut edges should be "buttered" with a high-alloy weld overlay.

2. Chloride ion contamination induces high corrosion rates, e.g., above 3000 ppm in 42% acid.[6] This is a result of oxidation of chlorides to nascent chlorine (as occurs in aqua regia), so that the effect diminishes with time as the chlorine is consumed or weathered off and the stainless repassivates.

3. Fluoride ions form hydrofluoric acid, and the resulting nitric-HF mixture becomes less oxidizing in nature, the fluoride ion resisting oxidation. (Conventional nitric-HF pickling mixtures reduce oxide films and embedded iron oxides, while causing active corrosion simultaneously with intergranular corrosion of sensitized areas.)

4. Vapor-phase corrosion is a potential problem for stainless steel tanks handling concentrations greater than 90%.

Storage and Handling

The following materials of construction are commonly used for storage and handling and their use is thought to constitute good engineering practice.

Tanks	Type 304L (S30403); aluminum greater than 93%
Tank trucks	S30403
Railroad cars	S30403
Piping	S30403
Valves	ACI CF3 (J92700); CF3M (J92800) acceptable
Pumps	ACI CF3 or titanium
Gaskets	Spiral-wound stainless/PTFE

References

1. R. R. Kirchheiner et al., "Increasing the Lifetime of Nitric Acid Equipment Using Improved Stainless Steels and a Nickel Alloy," *Materials Performance*, September 1989.
2. D. J. Newman and R. Miller, "Making Nitric Acid in All-Stainless Plants," *Chemical Engineering*, July 31, 1967.
3. C. P. Dillon, "Corrosion of Type 347 Stainless Steel and 1100 Aluminum in Strong Nitric and Mixed Nitric-Sulfuric Acids," *Corrosion*, vol. 12, no. 5, 1957.
4. R. D. Crooks, "Materials of Construction for Nitric Acid," *Process Industries Corrosion*, NACE, Houston, 1986.
5. "Handling Nitric Acid," *Chemical Engineering*, November 11, 1974.
6. M. W. Wilding and B. E. Paige, "Survey on Corrosion of Metals and Alloys in Solutions Containing Nitric Acid," ERDA ICP-1107, December 1976.

19

Oxygen

Introduction

Oxygen [boiling point (b.p.) −183°C; 90.2°K] is produced by liquefaction and fractional distillation of air at cryogenic temperatures, coproducing nitrogen (b.p. −196°C; 77.4°K) and small amounts of inert gases (i.e., argon, neon, krypton, and xenon).

The process demands metals and alloys immune to low-temperature embrittlement. Many iron-base alloys become sensitive to impact failure at temperatures below a critical value known as the *nil-ductility transition temperature.*

A powerful oxidizing agent, oxygen must not be allowed to contact most organic materials, since fire or explosion may result. In addition, practically all metals will burn in oxygen at ambient or higher pressures.

Oxygen may be produced on site for in-plant use, but pipeline plants can serve an industrial area and merchant plants provide the product in bulk shipment as a liquid or gas.

There are six grades of gaseous oxygen (A through F), varying from 99.0 to 99.995% purity and containing various kinds and amounts of minor impurities. Similarly, there are four grades of liquid oxygen (A at 99.0% purity and B, C, and D at 99.5%). Major concerns are inert gases (notably argon), odor, water, and total hydrocarbon content (THC).

Primary uses for oxygen are steel-making [e.g., the argon-oxygen decarbonization (AOD) process], nonferrous metallurgy (leaching, smelting), chemical processes (oxidation of hydrocarbons and other materials for the manufacture of ethylene oxide, ethanol, methanol, and acetic acid), and medical applications. Oxygen is also used in high-performance burners and to accelerate oxidation of sewage in waste-treatment plants.

Process

Ambient air intake is cooled and cleansed of water vapor and carbon dioxide (e.g., by freeze-out reversing exchangers or by molecular sieve absorbers). Solid adsorption agents are used to remove traces of other impurities. (Silica gel is particularly effective in removing traces of acetylene, which was responsible for many explosions in older plants.). Liquid oxygen and liquid nitrogen are separated by distillation (Fig. 19.1). The double-column rectifier has a connecting heat exchanger which serves as a condenser for the lower column oxygen and a reboiler for the nitrogen in the upper column.

Small-scale units produce oxygen by preferential absorption of nitrogen on zeolite resins (pressure-swing absorption). The product is about 90% oxygen plus argon and parts per million of water and carbon dioxide.

Materials

Aluminum alloys

Aluminum and its alloys are suitable, having good oxidation resistance and excellent low-temperature properties, and are widely used

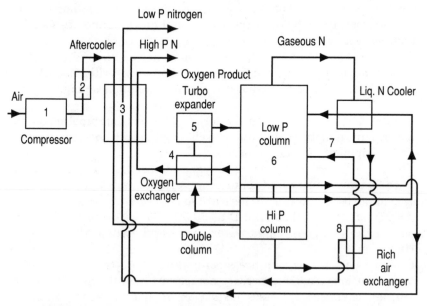

1.–8. S30400

Note: When a liquid oxygen vaporizer is employed, Aluminum A96061 is used on ambient air or S30403/steel on steam operation.

Figure 19.1 Oxygen (cryogenic process).

in oxygen plants. However, aluminum alloys have been known to ignite in high-purity oxygen, with catastrophic consequences. Consequently, many suppliers have desisted from using aluminum alloys for pressure vessels, pumps, valves, and other components in pure oxygen service.

Steel and cast iron

Steel at low temperature and cast-iron alloys have no application in this process because of brittle characteristics. Steel is used for ambient temperature storage of oxygen gas in cylinders. The 9% nickel steel is sufficiently ductile at cryogenic temperatures to be widely used for pressure vessels in this service.

Stainless steels

Austenitic stainless steels are suitable because of good oxidation resistance, cleanliness, and excellent low-temperature mechanical properties. Low-carbon grades are conventionally used, but are not required.[1] Thermal stress relief in the 840° to 950°C (1550° to 1750°F) range does not cause a significant reduction in impact properties, but is seldom called for in cryogenic applications.

Copper and its alloys

These have good low-temperature impact properties. Silicon bronzes have been used but have been replaced with austenitic stainless steels because of hot-short cracking propensities in welding. Aluminum bronzes have ignited in pump applications and are generally not used in liquid oxygen (LOX) service. However, other copper alloys possess excellent resistance to ignition, and copper or red brass piping systems with brass valves are widely used in pressurized piping systems for pure oxygen.

Nickel alloys

Nickel alloys are suitable. The nickel-copper alloy 400 (N04400) is widely used in difficult applications where ignition from high-temperature exposure or extreme mechanical damage is possible. This alloy has exceptional resistance to ignition, as well as good strength and toughness in both cryogenic and moderately high temperature applications.

Reactive metals

Titanium can react violently with oxygen if the surface oxide film is disturbed by high velocity or impact and is not suitable for this service. Zirconium is even more sensitive to impact. Tantalum and zirconium have no application in this service.

Precious metals

The precious metals have no application in oxygen service.

Other metals and alloys

Lead has no application in oxygen service. Tin is embrittled at cryogenic temperatures.

Nonmetallic materials

Organic. Most organic materials must be kept away from pure oxygen because of combustibility. Fluorocarbons, however, are used for valve seats and similar applications because of resistance to both low-temperature embrittlement and oxidation.

Inorganic. Ceramic materials are unaffected, but find no reported application in oxygen service.

Pitfalls

Problems with oxygen plants are associated with low-temperature impact properties, mechanical ignition of metals, and with inadvertent exposure to flammable materials.

Aluminum alloys have been a major concern because of an apparent sensitivity to mechanical ignition at low temperatures. A Compressed Gas Association (CGA) standard, G4.4, covers velocity limits for gaseous oxygen in carbon steel.

ASTM Standard G93 covers cleaning of equipment to remove organic materials and other potential ignition sources before service. Oxygen leaks contacting floor sweepings have caused explosions in plant.

Storage and Handling

The following materials of construction are commonly used for handling and storage, and their use is thought to constitute good engineering practice:

Tanks	Carbon steel (ambient temperature only)
Tank trucks	Tube trailers: steel; cryogenic trailers: 9% nickel steel or S30403

Railroad cars	Carbon steel (not for LOX)
Piping	Stainless steel, copper C12200, or steel (ambient temperature only)
Valves	CF8M, bronze, CF8
Pumps	CF8M, bronze, CF8
Gaskets	Spiral-wound PTFE/stainless steel

Anyone designing or using materials in oxygen service should be familiar with appropriate literature. The following references are specifically recommended as background information [from American Society for Testing and Materials (ASTM), Philadelphia, Pa. 19103]:

STP no. 812	"Flammability and Sensitivity of Materials in Oxygen-Enriched Atmospheres," vol. 1
STP no. 910	"Flammability and Sensitivity of Materials in Oxygen-Enriched Atmospheres," vol. 2
STP no. 986	"Flammability and Sensitivity of Materials in Oxygen-Enriched Atmospheres," vol. 3
STP no. 1040	"Flammability and Sensitivity of Materials in Oxygen-Enriched Atmospheres," vol. 4
G 63-87	"Standard Guide for Evaluating Nonmetallic Materials for Oxygen Service"
G 88-84	"Standard Guide for Designing Systems for Oxygen Service"
G 93-88	"Standard Practice for Cleaning Methods for Materials and Equipment Used in Oxygen-Rich Environments"
G 94-88	"Standard Guide for Evaluating Metals for Oxygen Service"

The following document is available from the Compressed Gas Association, Arlington, Va.: CGA G-4.4, "Industrial Practices for Gaseous Oxygen Transmission and Distribution Systems."

References

1. F. W. Bennett and C. P. Dillon, "Impact Properties of 18-8 Stainless Steels for Cryogenic Service," paper no. 65-WA/Met, American Society of Mechanical Engineers (ASME) Meeting, Chicago, 1965.

20

Phosphoric Acid

Introduction

Phosphoric acid is a moderately strong inorganic acid of a reducing nature, a syrupy liquid. It may be produced directly by absorption of phosphorus pentoxide in water (in which case there are few contaminants) or as a crude, contaminated product from the digestion of phosphate rock with sulfuric acid. In the latter case, oxidizing and reducing contaminants profoundly affect the corrosion characteristics.

Process

Electric furnace (Fig. 20.1)

Phosphate rock is nodulized and agglomerated by additions of silica and coke prior to reaction to produce elemental phosphorus. The liquid white phosphorus (P_4) is stored under water in steel tanks, corrosion being controlled by soda ash additions to neutralize the "phossy water."[1]

The liquid phosphorus is atomized with air and converted to phosphorus pentoxide (P_2O_5) in a graphite-lined combustor, cooled and absorbed in high-pressure water in a carbon-lined or S31603 absorber to form 75 to 85% phosphoric acid.[2]

Wet process (Fig. 20.2)

In the wet process, concentrated sulfuric acid is diluted with phosphoric acid and added to ground phosphate rock. The resultant mixture contains about 28% phosphoric (20% P_2O_5) and 21% sulfuric, plus fluorides (largely as hydrofluosilicic acid), chlorides, and metal cations (e.g., Fe^{+++}). The number and kind of contaminants, which profoundly influence the corrosion characteristics of wet-process phospho-

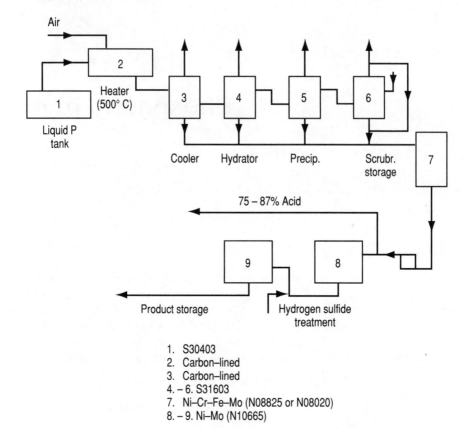

Figure 20.1   Phosphoric acid (electric furnace).

ric acid, vary from one geographic locale to another. Deposits of gypsum (calcium sulfate), formed from calcium phosphate, further complicate the corrosion process.

Rubber- or brick-lined digesters are employed, with high-alloy metallic components (e.g., N08028, N06985, N06030). After filtration to remove precipitated salts, the product acid concentration is roughly 35 to 44%.

Stronger concentrations of 58 to 75%, required for fertilizer applications, are obtained by vacuum evaporation or submerged combustion. Product acid may have iron and vanadium removed by treatment with potassium ferricyanide. Food-grade acid is made by precipitating fluorides, calcium, iron, aluminum, and sulfates. Arsenic can be precipitated by sodium sulfide. Fluoride levels may be reduced to less than 50 ppm by blowing with superheated steam or by reaction with sodium silicate.

Superphosphoric acid of 97 to 99% concentration is produced by fur-

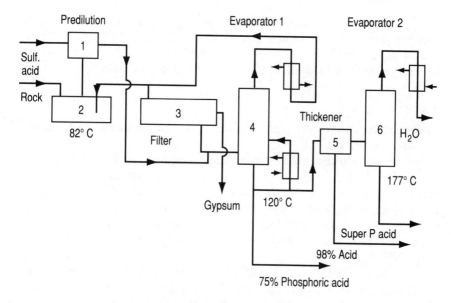

Figure 20.2   Wet-process phosphoric acid.

1. Ni–Cr–Fe–Mo (N08825 or N08020)
2. Brick–lined steel
3. – 4. Ni–Cr–Fe 6 Mo (N08028, N08026, N06007)
5. – 6. Ni–Cr–Fe–Mo (N08825 or N08020)

ther evaporation. It has a less corrosive nature than concentrated phosphoric acid and permits more economical transportation.

Materials

Aluminum alloys

Aluminum is resistant to phosphoric acid up to about 20% concentration to 65°C (150°F). It has no application in concentrated acid. Further, it would be adversely affected by any heavy metal ion or halide contamination. Dilute solutions inhibited with chromates have been used for cleaning and as a prepainting treatment.

Steel and cast iron

Steel forms a moderately protective phosphate film in acid above 70% concentration, and resistance may be further enhanced by arsenic salts. It is not usually employed in modern practice, because of iron contamination concerns.

Unalloyed cast iron has similar resistance in strong acid, but the effects of halides, sulfuric acid, and other contaminants must be carefully evaluated. Austenitic nickel-irons (e.g., F43000) are resistant to slightly above room temperature, but even the 14% silicon iron (F47003), nominally resistant in all concentrations to the atmospheric boiling point, can be attacked by fluoride contamination.[3]

Stainless steels

The straight chromium grades of stainless steel find no practical application in phosphoric acid.

Performance of nonmolybdenum grades of austenitic stainless steel (e.g., S30403) tends to be erratic. Nominally resistant to the boiling point to about 20% concentration and to about 65°C (150°F) in the 60 to 85% range, they are very sensitive to contamination effects. Traces of oxidizing metal cations (e.g., Cu^{++}) and oxidants (e.g., nitrates, nitrites) are beneficial, but halides (particularly chlorides) are inimical (see "Pitfalls" below).

Types 316L (S31603) and 317L (S31703) resist uncontaminated acid in all concentrations to about 80°C (175°F). However, velocities above about 1 m/s (3 ft/s) can cause erosion-corrosion in valves and piping.

It is precisely because of the contaminants in wet process acid that certain high-alloy, high-performance stainless steels were developed. Alloy 28 (N08028), a 27 Cr-4 Mo grade, and N08904 (20 Cr-25 Ni-4 Mo) have been successfully used in lieu of S31603, but N06985 and N06030 are usually preferred (see "Nickel and its alloys," below).

Copper and its alloys

Copper and its high-strength alloys (e.g., silicon bronze and aluminum bronze), as well as C70600 heat-exchanger tubes, have been successfully used in unaerated acid in the absence of other oxidizing agents (i.e., Fe^{+++}, Cu^{++}). They have been the material of choice in some high-temperature organic syntheses under reducing conditions (e.g., ethylene hydration).

Nickel and its alloys

Of the chromium-free alloys, N02200 and N04400 are of small use in phosphoric acid, being less resistant than copper alloys under reducing conditions and attacked under oxidizing conditions. The molybdenum grade, N10665, will resist *pure* acid of all concentrations to about 65°C (150°F) and up to 50% to the atmospheric boiling point. It is, however, rapidly corroded by oxidizing contaminants (e.g., Fe^{+++}).

Of primary interest is the specialty alloy N06007 and its newer variants, N06985 and N06030. These were developed especially for wet-process phosphoric acid.

The nickel-chromium-molybdenum grades (N06625, N10276, N06455, N06022) offer no advantage in pure acid. In contaminated acid, they are excellent in the presence of oxidizing cations alone, but are quite often not as resistant as the N06985, N08028, etc. in the presence of fluoride/chloride/sulfate ions as well.

Reactive metals

Because it is not an oxidizing acid (and because of frequent fluoride contamination), phosphoric tends to attack titanium, e.g., 20% at 85°C (185°F), although the titanium-palladium alloy is several times more resistant than unalloyed titanium. Fluorides, however, are a powerful negative influence.

Zirconium resists pure acid to about 50% above the atmospheric boiling point, but corrosion rates increase sharply above about 60% at 100°C. Fluorides increase corrosion dramatically.

Tantalum will resist all concentrations of phosphoric acid to about 190°C (375°F) *if* there is not more than a small (less than 10 ppm) fluoride contamination.

Noble metals

Gold and platinum will resist all concentrations of pure acid to the atmospheric boiling point.

Silver corrosion rates increase from less than 1 mpy in 15% acid at the atmospheric boiling point to 2 mpy in 60% and 7 mpy in 85% up to 250°C (480°F).

Other metals

Lead usually shows less than 20 mpy up to about 93°C (200°F). Resistance might be enhanced by sulfates and somewhat diminished by chloride and fluoride contamination.

Nonmetals

Organic materials. In the absence of solvent contamination, conventional plastics (e.g., PVC, PE, PP, epoxy, phenolic) resist all concentrations to the inherent temperature limit of the plastic. Troweled epoxy coatings are used to protect concrete tank pads from acid spillage. The fluorinated plastics are resistant to higher temperatures. Glass-reinforced plastics must have the glass component shielded from direct contact with the acid (see "Inorganic Materials" below).

Historically, wood has been used to contain some dilute phosphoric acid solutions, but stronger acid (greater than 10%) attacks wood by hydrolysis; plastic linings would be required.

Inorganic materials. Carbon and graphite are inert in all concentrations of acid, with or without contaminants, to at least 350°C (660°F). A film of phosphate protects against oxidation. Some impervious graphite *products* (e.g., tube and shell heat exchangers) are limited to about 170°C (340°F) by the constraints of cemented connections.

Glass is useful only at ambient temperature (even without fluoride contamination), but stoneware is resistant to about 80°C (175°F) and silica to 200°C (390°F).

Pitfalls

Attempts to store concentrated 85% phosphoric acid in S30403 tanks may have disastrous results, even at ambient temperature. Contamination with as little as 150 ppm chloride ion has evolved hydrogen within a tank, resulting in an explosion when the gas was ignited by an electrical liquid-level sensing device.

Storage and Handling

Following is a list of materials of construction for various items of equipment. Use of these materials is considered good engineering practice for the handling and storage of uncontaminated concentrated acid at ambient temperature, with minimum risk.

Tanks	FRP or S30403
Piping	FRP or S31603
Valves	CF3M
Pumps	CF3M
Gaskets	Elastomeric or flexible graphite

References

1. J. A. Lee, *Materials of Construction for Chemical Process Industries*, McGraw-Hill, New York, 1950.
2. "Corrosion Resistance of Nickel-Containing Alloys in Phosphoric Acid," *Corrosion Engineering Bulletin*, CEB-4, Inco Alloys International, 1966.
3. *Metals Handbook*, 9th ed., vol. 13, *Corrosion*, ASM International, Metals Park, Ohio, 1987.

Chapter

21

Sulfuric Acid
and Oleum

Introduction

Sulfuric acid is the largest tonnage inorganic chemical manufactured and is probably the most important, being a starting point for many other products.

Concentrated sulfuric acid, as an industrial product, is manufactured and shipped as either 93 or 98.5% acid. (The CP grade of 95.5% is distilled from the 93% commercial product in glass-lined stills, and shipped as a special product in glass bottles.) Oleum is fuming sulfuric acid, containing excess uncombined sulfur trioxide, and is shipped as 20 to 65% oleum.

Process

The basic reaction for manufacture of sulfuric acid by the contact process is the combustion of sulfur with dry air to sulfur dioxide, catalytic conversion to sulfur trioxide, and absorption of the product in strong acid. The final products are 98.5% acid and oleum, while 93% acid is produced by dilution in the air-drying tower.

The common basic materials of construction for the contact process are shown in Fig. 21.1.

Molten sulfur is burned in a brick-lined furnace with air which has been dried in a brick-lined steel tower with 98% acid, the heat of dilution to 93% concentration being removed in anodically protected type 316L (S31603) coolers. The hot SO_3 is cooled in a 2½ Cr-1 Mo (K31545) alloy steel waste-heat boiler and enters the converter (formerly brick-lined steel, currently type 304L preferred). A series of heat exchangers (S31003 superheaters and waste-heat boilers) controls the reaction temperature.

The SO_3 is absorbed in 98% acid to make oleum, which is diluted

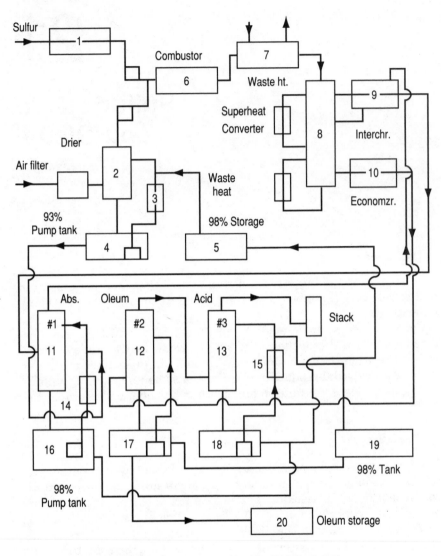

Figure 21.1 Double-contact sulfuric acid/oleum.

1., 3. S31603
2., 4., 5. Brick-lined steel
6. 6% Mo (S31254)
7. Carbon steel
8. S30403
9. - 10. 25 Cr-20 Ni (S31003)
11. - 13. 5 Si Stainless or
            brick-lined steel
14. - 15. S31603

16. - 19. Brick-lined steel
20. S30403

stepwise to 98.5% product (and to 93% in air drying) in brick-lined steel (or high-silicon stainless steel) towers. Pumps may be a high-alloy material (e.g., Lewmet 55) to resist abrasion as well as erosion-corrosion; otherwise they are usually cast alloy 20 (CN7M;N08007). (*Note:* Raw materials other than sulfur may be employed, such as waste acid or sulfidic ores. In such cases, the contact process is essentially the same after suitable drying of the sulfur dioxide and removal of abrasive particulate matter.)

Materials

Cast iron and steel

Ductile cast iron may be used for piping in concentrated acid (but not oleum) up to about 2 m/s (6 ft/s) with reasonable life. Carbon steel is used for storage tanks and vessels up to about 40°C (105°F). It should *not* be used for piping unless the velocity is kept below about 0.6 m/s (2 ft/s).[1]

Process iron (a specially manufactured cast iron of controlled microstructure) may be used in oleum and strong acid pumps.

Silicon cast iron (F47003) has phenomenal resistance to sulfuric acid up to 100% at the boiling point; it is unsuitable for oleum because of attack by sulfur trioxide.[2]

Stainless steels

The conventional 18-8 stainless steels (e.g., S30400, S31600, and their low-carbon variants) are used in strong acid (as the next material above cast iron) and oleum, to resist slightly elevated temperatures while handling higher-velocity flow. The type 316 (S31600) is *less* resistant than type 304 above about 93% acid, although it can be electrochemically protected (i.e., by *anodic protection*). Type 304 will resist 93% acid to about 40°C (105°F) and 98% acid to about 70°C (160°F), and has even higher temperature limits in a narrow range between 97.5 to 99.5% concentration.

Higher grades of conventional stainless, e.g., alloy 20 in the cast form, CN7M (N08007), are used for better erosion resistance (e.g., in motor valves and pumps). High-silicon grades are available for even better abrasion resistance (e.g., 17 Cr-17 Ni-5 Si) and higher temperature capabilities (e.g., to about 130°C).

Nickel-based alloys

Of the chromium-free grades, only the high-molybdenum alloy B-2 (N10665) has any potential application in concentrated sulfuric acid.

It will resist pure acid to the boiling point at 70% and to about 110°C (230°F) in 100% concentration. However, in manufacture of sulfuric acid, small amounts of nitric acid are formed by oxidation of the nitrogen in air, and alloy B-2 is *not* resistant in the product before refining to a meticulous grade (e.g., water-white or electrolytic).

In chromium-bearing alloys, alloy C276 (N10276) is useful under oxidizing conditions and resists 93 and 98.5% acid to about 90°C (195°F). However, the *chloride* content of the acid must be less than 40 ppm, or rates are increased by at least an order of magnitude.

Pitfalls

Steel

Velocity, turbulence, or impingement effects disturb the otherwise protective ferrous sulfate film and permit catastrophic rates of attack. Nitric acid contamination is inhibitive between 0.5 and 2%, but is otherwise harmful, oxidizing or solubilizing the iron sulfate film.

Cast iron

Regular gray (graphitic) cast iron can experience brittle failure of a catastrophic nature if inadvertently exposed to fuming sulfuric acid (e.g., where process iron was intended to be used).

18-8 stainless steel

Molybdenum additions reduce corrosion resistance above 93% acid concentration, and type 316 is not an acceptable substitute for type 304, as it can be in other environments where type 304 might be the alloy of choice. Chloride contamination *on the metal* (e.g., as from shipment or storage in marine atmospheres) can cause rapid failure by in-situ formation of HCl, unless the equipment is prewashed and dried.

Storage and Handling

Use of the following selected materials of construction is thought to constitute good engineering practice for handling and storage up to about 40°C (105°F).

Tanks	Carbon steel with type 304L bottom outlet nozzles. [Baked phenolic-coated or of all-stainless steel construction if iron contamination cannot be tolerated; steel tanks with anodic protection (AP) may be economical above about 1000 ton capacity.]

Tank trucks	Type 304L or baked phenolic-coated steel (plain carbon steel for oleum).
Railroad cars	Baked phenolic-coated steel for acid (carbon steel where iron pickup is permitted or for oleum).
Piping	Type 304 or 304L below 3 in, ductile cast iron above (except in oleum); polypropylene-lined steel to 8 in maximum in 93% acid only. (*Note:* PP may suffer chemical stress cracking (CSC) in dilute acids, and is attacked above 93%.)
Valves	CF8M or CF3M (fluorocarbon-lined for concentrated acid only); CN7M or specialty castings (e.g., high-silicon grades) for throttling valves.
Pumps	CN7M for acid; CK-20 for oleum.
Gaskets	Spiral-wound type 304/PTFE.

References

1. S. K. Brubaker, "Materials of Construction for Sulfuric Acid," *Process Industries Corrosion—The Theory and Practice*, NACE, Houston, 1986.
2. C. P. Dillon, "Corrosion of Type 347 Stainless Steel and 1100 Aluminum in Strong Nitric and Mixed Nitric-Sulfuric Acids," *Corrosion*, vol. 12, no. 5, 1957.

Part

3

Inorganic Processes

In this section, the manufacture of specific inorganic chemicals is described.
The general format for each chapter covers:

- *Introductory comments*

- *The process flow diagram (where appropriate) and descriptive text*

- *Discussion of relevant materials (Note:* to avoid needless repetition, inappropriate materials of construction are omitted in the individual chapters)

- *Pitfalls (in application of specific classes of appropriate materials in the specific process)*

- *Storage and handling (material choices thought to constitute good engineering practice for specific equipment)*

For more detailed information (e.g., on effects of impurities or problems associated with the use of a heavy chemical in other reactions or processes), the reader should consult the reference lists at the end of the chapters.

Chapter

22

Ammonium Chloride

Introduction

Ammonium chloride or sal ammoniac ($NH_4Cl \cdot nH_2O$) is an acid salt whose water solutions act like dilute hydrochloric acid. It is found in nature as a sublimation product of volcanic action. The corrosion characteristics are further complicated by the presence of the ammonium ion, which forms complexes with copper and nickel cations under oxidizing conditions.

Process

Ammonium chloride is a by-product from the production of sodium carbonate by the Solvay process, the reaction of ammonia, carbon dioxide, and sodium chloride producing sodium bicarbonate and ammonium chloride.

Another method is the double decomposition of ammonium sulfate with sodium chloride to produce sodium sulfate and ammonium chloride. The latter compound is in the filtrate and is recovered by crystallization.

Where sulfur dioxide is readily available, a similar reaction is effected by reacting ammonia, sulfur dioxide, and water with sodium chloride. This is in effect a double decomposition of ammonium sulfite, the ammonium chloride being recovered from the sodium sulfite mother liquor.

Direct reaction of ammonia with HCl is practical but not economically competitive with the other processes, as a rule. Nonmetallics or suitable alloys are used as materials of construction.

In one process, chlorine is reacted directly with ammonia to form CP ammonium chloride, using tantalum heat exchangers to concentrate the liquor ahead of the crystallizers. Evaporation of ammonium chloride solutions may be effected by using alloy 200 (N02200) heating coils or heat exchangers under nonoxidizing conditions.

Materials

Aluminum alloys

Aluminum is reported to be satisfactory for handling ammonium chloride, if the product is perfectly dry.[1] Pitting is observed at high humidities. Solutions of less than 1% have only mild action at ambient temperatures, but hotter or more concentrated solutions cause corrosion.[2]

Steel and cast iron

Ordinary steel and cast iron have been used to handle ammonium chloride where some corrosion and iron contamination can be tolerated.

High-silicon cast irons, alloyed to resist HCl (e.g., ASTM A-518, grade 2; a variant of F43007) have been used for hot solutions of medium concentration.

Stainless steels

Conventional 18-8 stainless steels are susceptible to pitting and SCC in ammonium chloride, depending on concentration, temperature, and pH. However, higher performance molybdenum-bearing grades (e.g., N08020 and the 6% Mo varieties) would be satisfactory.

Copper and its alloys

Copper and its alloys are generally unsatisfactory due to chromophoric corrosion products in the presence of air or oxidants, as well as the possibility of SCC by free ammonia.

Nickel alloys

Alloy 200 has been used for heating coils and heat-exchanger tubing, and nickel-clad steel has been used for hoppers, while alloy 400 (N04400) has been used in ammonium chloride flux tanks.

The chromium- and molybdenum-bearing grades (e.g., N06600, N06625, N06022) would have superior resistance to chloride pitting.

Reactive metals

Titanium should not be exposed to ammonium chloride solutions (which function as dilute HCl) unless there are ferric ions present to ensure passivity, but zirconium and tantalum should be resistant.

Precious metals

Silver, gold, and platinum have no reported application in this service. As with any ammoniacal solution, there are potential hazards in exposing silver under alkaline conditions, because of silver azide formation.

Other metals and alloys

Lead is attacked at 20 to 50 mpy by 0 to 10% solutions at ambient temperature.

Nonmetallic materials

Organic. Conventional plastics and elastomers resist ammonium chloride solutions up to about 80°C (175°F). Fluorinated plastics can be used to their normal pressure and temperature limitations.

Impervious graphite equipment is suitable.

Inorganic. Concrete is attacked by acidic ammonium chloride, but glassware and ceramicware are satisfactory.

Pitfalls

Pickup of atmospheric moisture will render ammonium chloride corrosive to aluminum and its alloys.

Zirconium might form pyrophoric corrosion products through intergranular attack (IGA) in the presence of moisture and iron contamination, in a manner analogous to behavior in HCl.

Acid-brick linings are chemically resistant, but the brickwork and cemented joints are susceptible to thermal shock in the crystallization process.

Storage and Handling

The solid product may be stored and shipped in polyethylene drums with a steel overpack.

The following materials of construction are commonly used for handling and storage of aqueous solutions, and their use is thought to constitute good engineering practice:

Tanks	Nickel-clad steel
Tank trucks	Nickel-clad steel
Railroad cars	Nickel-clad steel
Piping	Alloy 600
Valves	Nickel cast iron or CN7M
Pumps	CN7M
Gaskets	Graphite, spiral-wound PTFE/alloy 400, felted PTFE

References

1. J. A. Lee, *Materials of Construction for Chemical Process Industries*, McGraw-Hill, New York, 1950.
2. I. Mellan, *Corrosion Resistant Materials Handbook*, Noyes Data Corp., Park Ridge, N.J., 1976.

Ammonium Fluoride

Introduction

Pure ammonium fluoride (NH_4F), a white deliquescent crystalline salt, has characteristics of both hydrofluoric acid and the ammonium ion, analogous to ammonium chloride. In industrial practice, the corrosion characteristics will vary with the pH of water solutions.[1] Actually, the fluoride has a tendency to lose ammonia, forming the more stable ammonium bifluoride, which can be reconverted by reaction with a mole of aqueous ammonia if large quantities are needed. (The monofluoride is primarily a laboratory reagent.)

Ammonium bifluoride (NH_4HF_2) is a colorless, odorless crystal which is hygroscopic at humidities over 50%, melting at about 126°C (260°F). It is used for chemical dissolution of siliceous materials and recalcitrant iron oxides (e.g, loose mill scale; ferric oxide stains in textiles).

Process

Anhydrous ammonium bifluoride is made by dehydrating solutions of ammonium fluoride and drying the crystalline product.

The 93% commercial grade, containing about 0.1% water, is made by a gas-phase reaction of anhydrous ammonia with anhydrous HF (1:2 ratio) and flaking the molten product.

Materials

Aluminum alloys

Aluminum alloys have no reported application in this service, suffering moderate attack at room temperature.[2] Galvanic effects from heavy metal ions would pose serious problems in aqueous solutions.

Steel and cast iron

Iron and steel are fairly resistant to ammonium fluoride above about 20% contained fluorine (39% NH_4F). Below that concentration, they are unsatisfactory, due to dissolution of the protective iron fluoride films.

Stainless steels

Molybdenum-bearing grades, such as type 316L (S31603), are reported to be useful for aqueous solutions at room temperature. In hot acid solutions, as in stripping still tails, stainless steels are not serviceable.

Copper and its alloys

Copper and its alloys are not usually employed, because of the possible corrosion and SCC problems. However, they are sometimes employed (in lieu of the more expensive silver) under anaerobic conditions and when contamination is not objectionable.[3]

Nickel alloys

There are no reported uses of nickel-base alloys in this service. However, by analogy with copper alloys, as well as applications in ammonium chloride, they might be suitable under controlled conditions and after careful investigation.

Reactive metals

Titanium, zirconium, and tantalum are unsuitable, suffering embrittlement by the fluoride ion.

Precious metals

Silver is very satisfactory for aqueous solutions under acidic to neutral pH; silver evaporators permit large capacity and temperature range. The feed must be stripped to an acid range prior to entering the evaporator, as it is unsafe under ammoniacal conditions.

Other metals and alloys

By analogy with HF, lead and zinc would be nonresistant.

Nonmetallic materials

Organic. Conventional and fluorocarbon plastics may be employed within traditional temperature-pressure parameters.

Below about 95°C (200°F), rubber and elastomers may be employed for hose, pipe, and linings. Unlike HF, ammonium fluoride solutions do not harden rubber, and certain hard-rubber synthetics have been used up to about 130°C (265°F).

Impervious graphite heat exchangers and carbon-brick linings have been used up to at least 170°C (340°F) for evaporators, columns, and receivers or tanks.

Inorganic. The fluorides are incompatible with glass- and ceramic-ware.

Pitfalls

The possibility of silver azide formation in ammoniacal solutions poses a possibility of explosion.

Chloride ion contamination in aqueous solutions under acidic conditions could cause pitting, concentration-cell, and SCC problems with conventional 18-8 stainless steels.

Storage and Handling

Solid powdered product can be handled in coated steel containers or in polyethylene drums with a steel overpack.

The following materials of construction are commonly used for handling and storage of aqueous solutions, and their use is thought to constitute good engineering practice:

Tanks	Type 316L, epoxy-coated steel
Tank trucks	Type 316L
Railroad cars	Type 316L
Piping	PP-lined steel
Valves	CF3M
Pumps	CF3M or CN7M
Gaskets	Graphite fiber, PTFE/stainless, rubber

References

1. H. A. DePew, "Experiences in Handling of Ammonium Fluoride Solutions," *Trans. A.I.Ch.E.*, vol. 41, pp. 711–715, 1945.
2. I. Mellan, *Corrosion Resistant Materials Handbook*, Noyes Data Corp., Park Ridge, N.J., 1976.
3. J. A. Lee, *Materials of Construction for Chemical Process Industries*, McGraw-Hill, New York, 1950.

Ammonium Nitrate

Introduction

Ammonium nitrate (NH_4NO_3) is useful as fertilizer (usually as an ammoniated solution) and as an ingredient in certain explosives. Controlled decomposition produces the anaesthetic nitrous oxide. It does not occur in nature. As the solid product, a white crystalline solid, it is noncorrosive but aqueous solutions behave like dilute nitric acid unless ammoniated to an alkaline pH. The solutions have oxidizing characteristics due to the nitrate ion.

Process (Fig. 24.1)

Ammonium nitrate is manufactured by the reaction of ammonia with nitric acid in a type 304L (S30403) tank. The pH is automatically controlled, and the partially neutralized ammonium nitrate overflows to a second stainless steel tank for final adjustment to pH 6.4. This corresponds to an excess of about 50 ppm ammonia. The neutralized solution is pumped through stainless-steel piping to aluminum tanks which feed the type 304L evaporators operating under vacuum. The final product is "grained" to the crystalline form, which can be accomplished in steel, cast iron, or aluminum.

Materials

Aluminum alloys

Aluminum tanks, vessels, piping, and pumps or valves are eminently suitable for aqueous solutions. A93003 is more tolerant of acid conditions, but the 5000 and 6000 series are used up to 66°C (150°F). Copper-bearing alloys are not suitable, however. The nonsparking

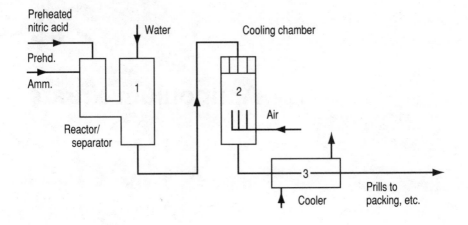

1. - 3. S30403

Figure 24.1   Ammonium nitrate.

characteristics of aluminum are insurance against accidental detonation.[1]

Steel and cast iron

Cast iron has traditionally been used for concentration, evaporation, and graining, so long as small amounts of corrosion and iron contamination are acceptable. Steel has likewise been employed in some plants but is attacked by dilute aqueous solutions of acid pH.

Silicon-cast-iron (F43007) pumps have sometimes been employed.

Stainless steels

Type 430 (S4300) has been widely used in ammonium nitrate manufacture but has been largely supplanted by the austenitic grades (e.g., S30403, S34700). The low-carbon or stabilized grades are required to prevent intergranular attack (IGA) corrosion under acid conditions (see Chap. 65). As in nitric acid, molybdenum-bearing grades offer no advantage over standard alloys.

Copper and its alloys

Copper and its alloys are unacceptable, being susceptible to corrosion by oxidizing acid solutions and by ammonium salts in the presence of oxidants. Further, copper corrosion products may form impact-sensitive compounds and lower the ignition point of some explosive formulations.[2]

Nickel alloys

Chromium-free nickel alloys have corrosion characteristics analogous to those of copper alloys and find no application in this service.

Chromium-bearing alloys (e.g., N06600) offer no advantage over austenitic stainless steels.

Reactive metals

The reactive metals would be resistant but are not economical or required in this service.

Precious metals

Silver is attacked by nitric-acid–type solutions and is hazardous in ammoniacal solutions. Gold and platinum are resistant but find no application in this service.

Other metals and alloys

Lead and zinc are attacked by ammonium nitrate solutions.

Nonmetallic materials

Organic. Various plastics are resistant to ammonium nitrate solutions, within their normal temperature and pressure limitations.

Impervious graphite construction is reportedly not resistant to acidic nitrate solutions, presumably because of attack on organic binders.

Inorganic. Glass- and ceramic-ware are resistant to aqueous solutions.

Concrete would be attacked by aqueous solutions of acidic nature and by wet ammonium nitrate (as in spillage of solid product).

Pitfalls

Ammonium nitrate is potentially explosive under certain ill-defined conditions; it was responsible for explosions in Nova Scotia and Europe and for the Texas City disaster of 1947.

Aluminum equipment may suffer galvanic corrosion by cementation if heavy metal ion contaminants are present. In the presence of free nitric acid in hot 83% solutions, preferential weld corrosion may occur.

The solid material is hygroscopic; unexpected corrosion of iron or steel will occur if there is pickup of moisture from the atmosphere.

The 18-8 stainless steels would be susceptible to SCC in the event of chloride contamination of aqueous solutions. Intergranular attack is a potential problem in the event of carbon contamination and sensitization of low-carbon and stabilized grades.

Storage and Handling

Solid ammonium nitrate is highly hygroscopic and cakes readily. It is usually shipped in moistureproof paper bags. Polyethylene drums would also be suitable.

The following materials of construction are commonly used for handling and storage of aqueous solutions, and their use is thought to constitute good engineering practice:

Tanks	Aluminum (3000, 5000 series) alloys
Tank trucks	Aluminum alloys
Railroad cars	Aluminum alloys
Piping	Aluminum alloys or S30403
Valves	CF3 or CF3M; copper-free aluminum
Pumps	CF3 or CF3M; copper-free aluminum
Gaskets	Spiral-wound PTFE/stainless steel

References

1. "Aluminum in the Chemical Industries," Aluminum Corporation of America, 1944.
2. J. A. Lee, *Materials of Construction for Chemical Process Industries*, McGraw-Hill, New York, 1950.

Chapter

25

Ammonium Phosphate

Introduction

Ammonium phosphate is produced in two forms; monoammonium phosphate ($NH_4H_2PO_4$ or MAP) and diammonium phosphate [$(NH_4)_2HPO_4$ or DAP]. MAP is used as a fire retardant for wood, paper, and cloth, but both MAP and DAP are used principally as fertilizers. The mono form has a pH of about 4.4 and is potentially corrosive, while the diammonium form is much less so, having a pH of about 8.

Process (Fig. 25.1)

A metered quantity of phosphoric acid, with or without sulfuric acid, is neutralized with anhydrous ammonia in agitated reaction tanks (brick-lined, with 6% Mo high-performance stainless agitators of undisclosed alloy designation). The final pH adjustment is effected in the second tank when the di-salt is being produced. The product is a thick slurry, to which finished product is recirculated—sometimes with additional inert material (gypsum and/or potassium sulfate or chloride) added between the reaction system and the blunger (a sloping device of S30403 construction for fluidization in a double-shaft, through-paddle mixer).

The granulated wet product from the blunger flows through a chute to a steel rotary cocurrent drier heated with hot air and combustion gases. The dried product is sifted over vibrating screens to obtain finished product of 8 to 10 mesh. Larger particles are fed to a pulverizer and the fines are recirculated to the blunger. Cyclone separators are used to remove dust from the recirculated air. Fumes are removed in an FRP or high-alloy stainless water scrubber, as required (the mono-salt at pH 4.4 has little tendency to give off ammonia, but the di-salt

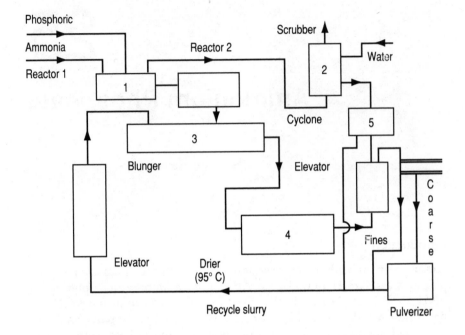

1. Brick-lined steel with Ni-Cr-Fe-6 Mo (N08028) agitator
2. FRP (bis-phenol)
3. - 5. Carbon steel

Figure 25.1   Ammonium phosphate.

is thermally unstable). The weakly acidic water is returned in polypropylene-lined steel pipe to the reaction system.

Materials

Aluminum alloys

Aluminum and its alloys may be used up to 28% concentration to about 60°C (140°F), provided some corrosion can be tolerated.[1] At elevated temperature, rates of more than 50 mpy may be encountered in 10% ammonium phosphate. Aluminum would be further attacked by any heavy metal ion or halide contamination, as when wet-process phosphoric acid is used.

Cast iron and steel

Cast iron and steel are routinely employed downstream of the reactors (or blunger if the mono-salt is produced). All equipment between the blunger and the scrubber can be of iron or steel construction, since iron contamination is not a problem with fertilizer products.

Stainless steels

The standard S30403 alloy is used for ammonia feed to the reaction system. However, the agitators must be of high-alloy (e.g., N08028) or rubber-coated-N08904 construction to withstand wet-process phosphoric acid (most commonly used in this process).

Copper alloys

Copper and its alloys are not suitable for ammonium salts with simultaneous exposure to moisture and air.

Nickel alloys

Nickel-based alloys have no application in this process.

Reactive metals

Neither titanium nor zirconium find application in this process.

Noble metals

Noble metals have no application in ammonium phosphate.

Other metals

Lead is resistant to ammonium phosphate solutions to about 66°C (150°F), but is not currently used.

Nonmetals

Organic. FRP construction is suitable for the scrubbing system, while polypropylene-lined steel pipe is suitable for the weakly acid scrubber-water tails. As with other inorganic salts and their aqueous solutions, almost any thermoplastic or thermosetting resin formulation may be used, as desired, up to its normal limiting temperature.

Inorganic. In some plants, impervious graphite evaporator heaters have been used, but are susceptible to breakage when slurry deposits must be removed.

Pitfalls

The primary pitfalls in ammonium phosphate production arise from contaminants (chlorides, fluorides, etc. from wet-process acid) which can cause pitting or SCC of 18-8–type stainless steels, erosion-corrosion or abrasion from the handling of slurries, and unexpected corrosion of steel or cast iron in the event of incomplete neutralization of the product.

Storage and Handling

The dried ammonium phosphate products are handled in steel containers.

References

1. I. Mellan, *Corrosion Resistant Materials Handbook,* Noyes Data Corp., Park Ridge, N.J., 1976.

Ammonium Sulfate

Introduction

Ammonium sulfate [$(NH_4)_2SO_4$] is produced both as the dry product, a white crystalline solid which is noncorrosive if completely dry, and as an aqueous solution. The latter has the characteristics of dilute sulfuric acid, unless neutralized. Its major use is as fertilizer.

Process

There are a number of processes used in production of ammonium sulfate. In short, ammonia from any source is scrubbed with an ammonium sulfate solution containing an excess of free sulfuric acid (nominally 5 to 6%, although concentrations of twice that value can be encountered in saturators). The scrubber tails is crystallized, the product being centrifuged or filtered and dried; the recovered ammonium sulfate mother liquor is fortified with sulfuric acid additions and recirculated.

In the wet stages, equipment must be rubber-lined or otherwise constructed of corrosion-resistant materials to withstand the acidic solutions.

Materials

Aluminum alloys

Aluminum has been successfully used in chloride-free neutral solutions to about 70°C (160°F), no free sulfuric acid being present.

Steel and cast iron

Iron and steel are not suitable for ammonium sulfate solutions, because of the prevailing acidity, although cast-iron pumps have been used in neutral solutions.

Silicon cast iron (F47003) has been used for piping as well as valves and pumps.

Stainless steels

Type 316L (S31603) has been successfully used for dilute acid solutions for centrifugal screens, saturators, etc. and, in the cast form (CF3M), for saturator pumps.

Higher alloys (e.g., CN7M) are preferred to combat erosion-corrosion and/or higher acidities than can be tolerated by type 316L.

Copper and its alloys

Copper and bronze equipment (e.g., centrifugal extractors) can be utilized, provided there is neither free ammonia nor air or oxidizing agents present.[1]

Nickel alloys

Alloy 400 (N04400) has been used for heating coils, but may suffer accelerated attack because of erosion-corrosion or hot-wall effects (especially above about 93°C (200°F). Otherwise, performance has been good and alloy 400 construction has been used for acid conditions in saturators, scrubbers, mother-liquor tanks, settling tanks, centrifuges, vacuum filters, centrifuges, saturator pumps, and ancillary equipment.[2]

The chromium- and molybdenum-bearing grades (N06625, N06022, N10276) could be used under more severe conditions.

Reactive metals

Titanium would be unsatisfactory under acid conditions, but zirconium and tantalum can be used for severe conditions.

Precious metals

Silver is reportedly unattacked in ammonium sulfate solutions containing 5% sulfuric acid. Gold and platinum find no application in this service.

Other metals and alloys

Lead shows less than 20 mpy in ammonium sulfate solutions at ambient temperature.

Nonmetallic materials

Organic. Phenolic and vinyl resin coatings may be used, although corrosion at holidays would be a potential problem.

Assorted plastics and elastomers are suitable, depending on sulfuric acid concentration and the temperature-pressure parameters.

Impervious graphite is resistant, and has replaced alloy 400 in some heat-exchanger applications.

Inorganic. Glass- and ceramic-ware or acid-brick construction may be used as desired. Concrete is attacked both by acid conditions and by the sulfate bacillus effect.

Pitfalls

Chloride contamination by salts of heavy metals causes cementation attack of aluminum alloys.

Pyrophoric corrosion products could form from IGA of zirconium in dilute acid solutions with cupric or ferric ion contamination.

Storage and Handling

The dry product is packaged in multiwall paper bags and fiber or PE drums. It is also shipped in bulk via railroad cars.

The following materials of construction are commonly used for handling and storage of aqueous solutions, and their use is thought to constitute good engineering practice:

Tanks	Carbon steel
Tank trucks	Carbon steel
Railroad cars	Carbon steel
Piping	Carbon steel
Valves	CF3M
Pumps	CN7M
Gaskets	Rubber, graphite fiber, PTFE/stainless steel spiral-wound

References

1. F. L. LaQue, *Canadian Chem. Proc. Ind.*, 1938, p. 185.
2. J. A. Lee, *Materials of Construction for Chemical Process Industries*, McGraw-Hill, New York, 1950.

Chromic Acid

Introduction

Chromic oxide (the acid anhydride) is one of the oldest green pigments known. Although the oxide is substantially noncorrosive to iron and steel, the aqueous solutions which constitute chromic acid (H_2CrO_4) are a problem. A powerful oxidizing acid, chromic acid solutions may nevertheless be incompatible with stainless-steel construction, and even with high-silicon cast irons under some conditions.[1]

Process

In one process, sodium dichromate is added to 96% sulfuric acid in a steel tank and the mixture heated to about 200°C (390°F). At this reaction temperature, the mixture separates into a two-phase molten mass, the upper layer sodium bisulfate and the lower one chromic acid anhydride [melting point (m.p.) 197°C]. The chromium trioxide is drawn off to solidify in cast-iron pans, then crushed and packed as a solid product. If 20% oleum is substituted for the sulfuric acid, little external heat is required, liquid chromic acid being spontaneously produced. This reaction has been used for a continuous production process.

Materials

Aluminum alloys

A 1 to 10% solution is reported to have mild action up to 66°C (150°F) on aluminum.[2]

Steel and cast iron

Steel is compatible with dry chromic oxide, but attacked by chromic acid solutions. In intermediate steps (e.g., flaking operations), steel gives only limited service life because of acid attack combined with erosion or abrasion. At less than 4 mpy [0.1 (mm/a)], the following limits are observed: 40% at 20°C (68°F), 20% at 30°C (86°F), and 15% at 40°C (105°F).[3]

Cast iron is useful only in the evaporation step described above, and only when iron contamination is acceptable.

The high-silicon cast iron (F43007) is resistant to a wide range of pure and crude chromic acid at up to 40°C (105°F), as in pumps, valves, piping, etc., for the solutions encountered in chrome-plating operations. It is, however, corroded in some dissolution procedures.

Stainless steels

The 18-8 variety of stainless steels resist chromic acid at room temperature up to 30%.[3] However, rates well above 4 mpy (0.1 mm/a) are observed above 5% concentration at about 80°C (175°F). Because of its oxidizing nature, chromic acid is more aggressive to S31603 than to S30403.

High-alloy stainless castings, e.g., CN7M, have been successfully used in pumping various concentrations and temperatures. However, they cannot be directly used in chromic acid dissolution steps.

Copper and its alloys

Copper alloys find no application in this service.

Nickel alloys

Chromium-free alloys find no application in this service. Alloy C276 (N10276) will resist 10% acid to about 95°C (200°F).

Reactive metals

Titanium, zirconium, and tantalum show less than 2 mpy in 50% acid to about 100°C (212°F).[4,5]

Precious metals

Silver is resistant to all concentrations to the atmospheric boiling point.

Other metals and alloys

A lead alloy containing 7% tin has been used for handling chromic acid solutions. Lead resists 85% chromic acid to about 220°C (430°F), 93% to 150°C (300°F), and 95% acid at ambient temperatures. In dilute solutions (e.g., less than 5%), the rate increases but is still acceptable.

Tin resists 80% acid to about 100°C (212°F).

Magnesium is very resistant in the absence of chloride ions. A boiling 20% solution has been used to clean magnesium alloys without attacking the base metal.[5]

Nonmetallic materials

Organic. Chlorinated polyethers and polyesters, CPVC, PE, and PP resist up to 50% chromic acid to about 70°C (160°F). Fluorinated plastics are resistant to their usual temperature limits. Vinyl copolymer coatings have been successfully used for some dilute solutions.

Fluorinated elastomers are satisfactory to about 100°C (212°F), depending on specific formulation. Buna N is resistant to about 80°C (175°F).

Inorganic. Glass- and ceramic-ware are frequently used to resist chromic acid solutions. A special iron-free chemical stoneware tile set with acid-resistant cement has been used for tanks containing chromium-plating solutions.

Pitfalls

Halide contamination can adversely affect the corrosion resistance of aluminum and stainless alloys.

Storage and Handling

The solid chromic oxide can be shipped in steel drums.

The following materials of construction are commonly used for handling and storage of chromic acid solutions, and their use is thought to constitute good engineering practice:

Tanks	Type 304L (S30403)
Tank trucks	Type 304L
Railroad cars	Type 304L
Piping	Type 304L
Valves	CF3 or CF3M
Pumps	CN7M
Gaskets	Spiral-wound PTFE/stainless steel

References

1. J. A. Lee, *Materials of Construction for Chemical Process Industries*, McGraw-Hill, New York, 1950.
2. I. Mellan, *Corrosion Resistant Materials Handbook*, Noyes Data Corp., Park Ridge, N.J., 1976.

3. *Corrosion Tables—Stainless Steels,* Jernkntoret, Stockholm, 1987.
4. P. A. Schweitzer, "Corrosion Resistance Tables," 2d ed., Marcel Dekker, New York, 1986.
5. B. D. Craig, *Handbook of Corrosion Data,* ASM *International,* Metals Park, Ohio, 1989.

28

Sodium Hypochlorite

Introduction

Sodium hypochlorite (NaOCl) is used for bleaching and sterilizing operations as a convenient method for introducing active chlorine with simultaneous safety and ease of handling. NaOCl contains 47.65% available chlorine (i.e., a 25% solution is equivalent to about 12% chlorine). Concentrations up to 15 to 23% chlorine are encountered, while commercial bleach is about 5% NaOCl (2.4% chlorine). Bleach is usually diluted about tenfold (3 to 3.5 g/L of available chlorine). Corrosion characteristics will vary greatly with the concentration, the more resistant materials (e.g., Ni-Cr-Mo alloys, glass, rubber, and plastics) being required in concentrated solutions.

Trace metal contamination (e.g., nickel, cobalt, copper, and iron) causes decomposition.

Process

Sodium hypochlorite is made commercially by reacting chlorine with a solution of caustic soda (NaOH). Corrosion-resistant pipe (e.g., lead, nickel alloys, fluorinated plastic) is used to introduce liquid or gaseous chlorine into sodium hydroxide in stoneware or PVC-coated or -lined concrete tanks. The high-silicon cast irons (F43000) and the molybdenum-bearing variant (ASTM A-518, grade 2) have been used for pumps, valves, piping, and steam ejectors. Cooling coils should be titanium, although alloy C276 (N10276) has been used.

Materials

Aluminum alloys

Aluminum and its alloys are incompatible with sodium hypochlorite solutions.

Steel and cast iron

Steel and cast iron are corroded by sodium hypochlorite solutions, but calcium hypochlorite solutions are less corrosive.

Stainless steels

Reportedly, types 316L (S31603) and 317L (S31703) have been successfully used in dilute alkaline solutions up to about 0.3% available chlorine.[1]

High-performance grades (e.g., S31254, N08020, etc.) might be used to about 3% available chlorine but not in more concentrated solutions.

Copper and its alloys

Copper alloys generally find no application in hypochlorite solutions.

Nickel alloys

The chromium-free grades (N02200, N04400) have been successfully used to about 0.3% available chlorine in alternate alkaline and acid exposures such as cyclic textile bleaching. In more concentrated solutions, corrosion is severe.

The chromium-bearing alloy 600 is resistant to dilute solutions, but susceptible to pitting and concentration-cell attack. The nickel-chromium-molybdenum grades (N06626, N06022, N10276) are very resistant. Cast versions (e.g., CW-2M) are used for pumps and valves for making and handling bleach solutions.

Reactive metals

Titanium, zirconium, and tantalum are resistant to hypochlorite solutions. Heat exchangers and cooling coils are usually titanium.

Precious metals

There are no reported applications of gold or platinum in this service, but silver chlorination coils and sparger tubes have been used in some applications.

Other metals and alloys

Lead has been used in the past for introducing chlorine to the reaction mixture.

Nonmetallic materials

Organic. PVC and PVDC are resistant to sodium hypochlorite solutions to about 80°C (175°F), while fluorinated plastics are useful at higher temperatures. Chlorination vessels and reactors are often FRP (vinyl ester) with a thermoplastic liner. PE has been used for dilute bleach, while PP is susceptible to CSC.

Rubber-lined tanks, pipes, and fittings with synthetic rubber (e.g., EPDM, chlorobutyl) are successfully used.

Impervious graphite, with an epoxy-based binder, has been used with 25% hypochlorite to the atmospheric boiling point (phenolic binders are attacked by the hypochlorite).

Inorganic. Glass-lined steel has been used for 16% hypochlorite solutions at ambient temperature.

Pitfalls

The austenitic stainless steels are prone to pitting, crevice corrosion, and SCC in hypochlorite solutions.

Plastic additives (e.g., thixotropes, catalysts) are a possible source of objectionable metal ion contamination.[2]

Glassware is susceptible to attack at moderately elevated temperatures and/or high pH.

Concrete tanks have been used, but they are sometimes coated to minimize deterioration. A trowelable epoxy coating will resist 27% sodium hypochlorite at ambient temperatures, to 6% contained chlorine up to about 70°C (160°F).[3]

Storage and Handling

The following materials of construction are commonly used for handling and storage, and their use is thought to constitute good engineering practice:

Tanks	Rubber-lined steel
Tank trucks	Rubber-lined steel
Railroad cars	Rubber-lined steel
Piping	PVDC-lined steel

Valves	Chlorimet 3, Durichlor
Pumps	Chlorimet 3, titanium, FRP (vinyl)
Gaskets	Rubber, graphite fiber

References

1. J. A. Lee, *Materials of Construction for Chemical Process Industries*, McGraw-Hill, New York, 1950.
2. J. K. Nelson, "Materials of Construction for Alkalies and Hypochlorites," *Process Industries Corrosion—The Theory and Practice*, NACE, Houston, 1986.
3. I. Mellan, *Corrosion Resistant Materials Handbook*, Noyes Data Corp., Park Ridge, N.J., 1976.

Sulfur

Introduction

Elemental sulfur is an amorphous yellow solid used as a raw material for manufacture of sulfur dioxide, sulfuric acid, and related products. It is also used to formulate a sulfur cement for some corrosive services.

It occurs in nature both in the free state and as sulfidic ores of iron, zinc, and copper/iron ore (chalcopyrite). Although about 90% of sulfur production is used to manufacture sulfuric acid, it also finds process applications in the rubber, chemical, paper, and pharmaceutical industries.

Process

Sulfur has been mined manually in some areas. However, large amounts are produced by "mining" with hot water (the Frasch process). It is also produced by interaction of hydrogen sulfide and sulfur dioxide. By-product sulfur now constitutes the larger supply, being recovered from gas, petroleum, coal, and sulfidic ores.

In the Frasch process, hot water at 160°C (320°F) is pumped down a concentric pipe arrangement to melt the sulfur, which is then forced up to the surface by hydraulic pressure. Both fresh water and seawater have been used for this purpose.

Sulfur is also produced by open-pit mining, a portion being burned to provide heat to melt the remaining crude sulfur.

Hydrogen sulfide is converted to elemental sulfur by burning a portion to sulfur dioxide and interacting the gas streams in a modified Clauss process.[1]

A number of processes, not economically competitive, have been developed for producing elemental sulfur directly from pyritic ores.

Materials

Aluminum alloys

Aluminum and its alloys are resistant to both liquid and vapor forms to at least 410°C (770°F).[2] Aluminum prilling towers are in common use.

Metallized aluminum coatings have been used over a steel substrate to protect mild steel at temperatures as high as 440°C (825°F).

Steel and cast iron

Iron and steel give satisfactory service to about 175°C to 205°C (350° to 400°F) in the absence of air and moisture contamination. Any increase in temperature and moisture (with or without aeration) can cause excessive corrosion. Dissolved sulfur is analogous to dissolved oxygen in the corrosion process.

Stainless steels

Conventional ferritic (e.g., S43000) and 18-8 stainless steels are satisfactory up to about 205°C (400°F), but offer small advantage over steels.

Above the boiling point of 444°C (832°F), a 27% ferritic grade is somewhat more resistant than type 310 (S31000), although close to the "blue embrittlement" temperature of 474°C (885°F). Either grade is preferable to type 304L and its variants.

Copper and its alloys

Copper alloys are somewhat better than copper, but the entire family tends to be unsatisfactory due to formation of voluminous black sulfide corrosion products.

Nickel alloys

Alloys 200 (N02200) and 400 (N04400) are resistant up to about 300°C (575°F), if free access of air is prevented. Above that temperature, corrosion rates are greatly increased.

The chromium-bearing grade, alloy 600 (N06600) is an order of magnitude better.

Alloy C276 (N10276) is the most resistant of the nickel-base alloys.

Reactive metals

Titanium is unaffected by molten sulfur at 240°C (465°F). The reactive metals are not usually considered for this service.

Precious metals

Silver is tarnished by sulfur and sulfides.

Other metals and alloys

Magnesium is inert to molten or gaseous sulfur.

Chromium, usually used as chromium plate on a steel substrate, is resistant to both elemental sulfur and sulfides, alone and in aqueous solution.

Nonmetallic materials

Organic. PVC, PE, and PP are satisfactory below about 60°C (140°F). PVDF is suitable to 120°C (250°F), and other fluorinated plastics can be used to their normal temperature limitations.

Selected FRP materials will resist molten sulfur to about 120°C (250°F).

Elastomers tend to undergo continued vulcanization in molten sulfur, but selected fluorinated grades may be used to about 115°C (240°F). Chlorsulfonated polyethylene and neoprene may be used at ambient temperatures.

Furane, phenolic, and acid-resistant silicate mortars (but not sulfur-based cements) resist molten sulfur.

Inorganic. Ceramicware is reportedly resistant to sulfur, sulfides, and polysulfides to about 150°C (300°F).[3]

Pitfalls

Conditions of moisture, aeration, and temperature affect formation of sulfur-based acids, which can profoundly affect corrosion of various metals and alloys.

If steel piping is allowed to drain and cool after transporting molten sulfur, oxygenated sulfur-based acids will form and cause severe corrosion.

Hydrogen sulfide may be present in fumes from molten sulfur, creating problems of toxicity as well as fire or explosion if released under certain conditions.

Molten sulfur is itself combustible, and should be kept below about 150°C (300°F). It must be protected from open flames, electric sparks, and other sources of ignition. (An aluminum prilling tower burned down, apparently ignited by a thermite reaction between the aluminum vessel and rusted steel support members.)

Sulfur-contaminated nickel-base alloys are susceptible to liquid-metal cracking (LMC) by the nickel-sulfide eutectic during welding or other high-temperature operations (see Chap. 60).

Storage and Handling

About 95% of the sulfur handled in North America (and about 50% worldwide) is handled in the molten state.[1] This practice has the advantages of high purity, minimum losses during handling, and ease of transfer to pit or storage tank. Also, one avoids certain problems involved in handling solid sulfur, such as effects of acid formation from ingress of moisture due to changing climatic conditions, airborne sulfur contamination, and higher handling losses.

Molten sulfur requires careful temperature control to maintain a range between the melting point of about 115°C (238°F) and 160°C (320°F), above which there is *increasing* viscosity. In practice, the temperature should never be allowed to drop below 116°C (241°F) or rise above 150°C (300°F). The flash point is about 248° to 261°C (478° to 502°F).

The following materials of construction are commonly used for handling and storage, and their use is thought to constitute good engineering practice.

Tanks	Insulated carbon steel with steel heating coils
Tank trucks	Insulated carbon steel
Railroad cars	Insulated carbon steel, aluminum
Piping	Carbon steel, insulated and traced
Valves	Cast steel, insulated and traced; cast aluminum (A03560)
Pumps	Cast steel, insulated and traced
Gaskets	Flexible stainless steel, PTFE, graphite fiber

References

1. G. T. Austin, *Shreve's Chemical Process Industries*, 5th ed., McGraw-Hill, New York, 1985.
2. M. Krienberg, "Handling Molten Sulfur," *Chemical Engineering*, December 4, 1978.
3. J. A. Lee, *Materials of Construction for Chemical Process Industries*, McGraw-Hill, New York, 1950.

30

Sulfur Dioxide

Introduction

Sulfur dioxide (SO_2), the anhydride of sulfurous acid, is a colorless gas with a distinctive pungent odor. With a boiling point of $-10°C$ ($14°F$), it is stored under compression as a liquid product. When wet or in aqueous solution is has the characteristics of a reducing acid.

Process

The common process for manufacture of sulfur dioxide involves simple burning of sulfur (Fig. 30.1). It is also produced by burning other sulfur-containing materials (e.g., sour waste gases, sulfide ores).

The produced sulfur dioxide is cooled in steel or stainless-steel exchangers (depending on the cooling-water system) and passed to the acid-brick-lined absorbers to produce a dilute sulfurous acid (1 to 2%, depending on the temperature and SO_2 content of the entering gas). The gas enters a brick-lined steaming tower and then a cooler. It is then dried with 98% sulfuric acid in a tower, the overhead product being compressed, cooled, and stored in steel pressure vessels.

Materials

Aluminum alloys

Aluminum alloys are resistant to dry sulfur dioxide and to humidified SO_2 to allowable ASME code temperature of about $204°C$ ($400°F$). Mild action may occur in dilute solutions (in the absence of chlorides or heavy metal ions). Aluminum finds application in towers used for SO_2 treatment of corn products and in refrigerant systems using sulfur dioxide (where the metal's excellent low-temperature properties are also beneficial).

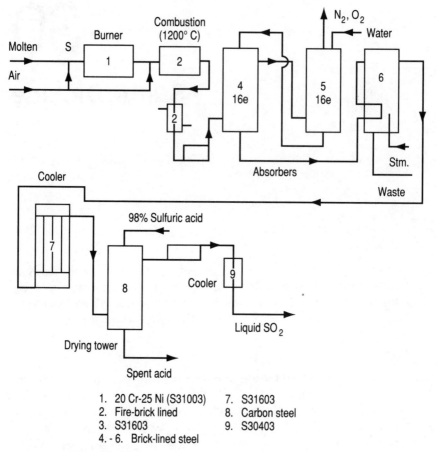

1. 20 Cr-25 Ni (S31003)   7. S31603
2. Fire-brick lined   8. Carbon steel
3. S31603   9. S30403
4. - 6. Brick-lined steel

Figure 30.1   Sulfur dioxide.

Steel and cast iron

Steel is suitable for *dry* sulfur dioxide only. Ordinary gray cast iron and steel are corroded by the wet gas and by sulfurous acid.

High-silicon cast iron (F47003) is *not* recommended for sulfurous acid, but austenitic nickel cast irons (F43000) have been used for gas manifolds and distributors carrying hot mixed gases from sulfur burners to coolers in sulfite pulp mills.[1]

Stainless steels

Non-molybdenum-bearing grades (e.g., S30403) are attacked by sulfurous acid at rates greater than 40 mpy.

Conventional stainless steels (e.g., S31603, S31703) are widely used for sulfurous acid solutions at ambient temperature and for dry gaseous SO_2-containing combustion gases. They are subject to crevice corrosion, however.

When traces of sulfuric acid are present, more highly alloyed stainless steels (N08020, CN7M) must be employed (e.g., in sulfurous acid tower circulating pumps). They are also used for SO_2-rich sulfuric acid, but require inhibition (e.g., with peroxide) at moderately elevated temperatures, e.g., 50% sulfuric acid at about 50°C (120°F).

In the simultaneous presence of chlorides, higher alloys are required. Type 317L (S31703) and high-performance alloys (e.g., S31254, N08904, N08367) are used as the temperature, concentration, and chloride content are increased (e.g., in pulp and paper digestion and in flue-gas scrubbers). Even with rates less than 1 mpy for N08020 and similar compositions, crevice corrosion can be a problem. The 6% Mo grades (e.g., S31254, N08367) are preferred.

Copper and its alloys

Copper alloys are resistant only to dry sulfur dioxide or wet gases under *noncondensing* conditions. In acid solutions, they are attacked at greater than 40 mpy, with the corrosion aggravated by dissolved oxygen. Attack is autocatalytic with accretion of cupric ions.

Nickel alloys

Alloys 200 and 400 (N02200, N04400) resist the dry gas and wet gas under noncondensing conditions, although superficial tarnishing may occur. The upper temperature limit is about 300°C (600°F), above which intergranular attack and sulfur embrittlement (LMC) may occur.

Alloy 600 (N06600) is much superior, resisting SO_2 to about 815°C (1500°F). However, alloys 200, 400, and 600 will not resist sulfurous acid solutions containing more than about 0.3% by weight.[1]

In sulfite solutions at elevated temperatures and in the presence of chlorides (e.g., in flue gas scrubbing units), N10665 is attacked at about 5 mpy. Ni-Cr-Mo alloys (N06625, N06022, N10276) are the ultimate in corrosion resistance.

Reactive metals

Titanium is reportedly unaffected by sulfurous acid, even with chloride contamination, to about 150°C (300°F).

Data on zirconium are unavailable.

Tantalum is completely resistant (except when sulfur trioxide is also present) but is rarely required. Tantalum equipment is sometimes employed in food, drug, and fine chemical applications where SO_2 is present in conjunction with other corrosive species.

Precious metals

There are no reported applications of silver, gold, or platinum in this service. Silver is attacked at red heat, but gold and platinum are not.[2]

Other metals and alloys

Lead has been widely used, especially in SO_2/H_2SO_4 services, and is useful to about 200°C (390°F) if properly supported.

Zinc is corroded, but magnesium and tin are resistant.

Nonmetallic materials

Organic. Common plastics and FRP will resist sulfurous solutions within normal parameters of temperature and pressure, but fluorocarbon plastics are preferred at elevated temperatures or where traces of sulfuric acid are present.

Rubber-lined equipment and hard-rubber components resist SO_2 and saturated sulfurous acid solutions to about 80°C (174°F). (Soft rubber is more permeable to sulfur dioxide, and linings may suffer blistering or disbondment.) Butyl rubber seems to be the only synthetic formulation offering any advantage over natural rubber.

Impervious graphite and carbon equipment may be used for all concentrations of sulfurous acid or its anhydride. A temperature limit of about 170°C (340°F) is assigned to cemented impervious graphite exchangers.

Inorganic. Glass-lined steel can be used to 260°C (500°F). Chemical stoneware can be used within its normal parameters of service, limited only by potential thermal shock.

Pitfalls

Crevice corrosion and concentration-cell effects are a potential problem with 18-8 types of stainless steels; higher-performance grades (e.g., N08904, S31254, N08367) are preferred especially in the simultaneous presence of halide contamination.

Weldments of molybdenum-bearing stainless steels (S31603) are subject to preferential corrosion, because of segregation during solidification, in chloride-contaminated sulfurous acid (e.g., pulp and paper processes, flue gas desulfurization [FGD] scrubbers). Welds should be made with an overmatching rod in such service (e.g., N08904 or N08367 to weld S31703; N06625 or N06022 to weld 6% molybdenum high-performance grades).

Storage and Handling

The following materials of construction are commonly used for handling and storage of the dry compressed gas, and their use is thought to constitute good engineering practice:

Tanks	Carbon steel
Tank trucks	Carbon steel
Railroad cars	Carbon steel
Piping	Carbon steel
Valves	Cast steel, cast iron
Pumps	Cast steel
Gaskets	Butyl rubber, spiral-wound PTFE/stainless steel

References

1. J. A. Lee, *Materials of Construction for Chemical Process Industries*, McGraw-Hill, New York, 1950.
2. B. D. Craig, *Handbook of Corrosion Data*, ASM International, Metals Park, Ohio, 1989.

Organic Syntheses

In this section, the manufacture of specific organic chemicals is described.
The general format for each chapter covers:

- *Introductory comments*

- *The process flow diagram and descriptive text*

- *Discussion of relevant materials (Note:* To avoid needless repetition, inappropriate materials of construction are omitted in the individual chapters)

- *Pitfalls (in application of specific classes of appropriate materials in the specific process)*

- *Storage and handling (material choices thought to constitute good engineering practice for specific equipment)*

For more detailed information (e.g., on effects of impurities or problems associated with the use of a heavy chemical in other reactions or processes), the reader should consult the references.

Acetaldehyde

Introduction

Acetaldehyde (CH_3CHO) is the next higher homolog of formaldehyde (Chap. 50). It is a colorless liquid with a pungent odor. It is used primarily as an intermediate for other organic chemicals (e.g., acetic acid and anhydride, peracetic acid, glyoxal, acetone, esters, and amines).

Process

Production from acetylene via the mercury-salt-catalyzed (e.g., mercuric sulfate) process is now obsolete because of environmental concerns and technological developments.

Modern processes are primarily (1) oxidation of ethylene and (2) oxidation of ethanol.

Ethylene oxidation

Both two-stage (Fig. 31.1) and one-stage (Fig. 31.2) processes are employed. Oxidation occurs over a mixed palladium chloride–cupric chloride catalyst.

In the two-stage process, ethylene is reacted with air in a titanium tubular reactor at about 125°C and 1.13 MPa (150 lb/in²). Crude acetaldehyde is flashed overhead, the catalyst solution being recycled. The product gases are water-scrubbed, and refined acetaldehyde is produced by distillation.

In the one-stage process, ethylene and oxygen (together with recycle gases) are reacted, the gases water-scrubbed, and the resulting solution distilled. In this process, refined acetaldehyde is a tails product, the tails gas being taken overhead for recycling to the reactor.

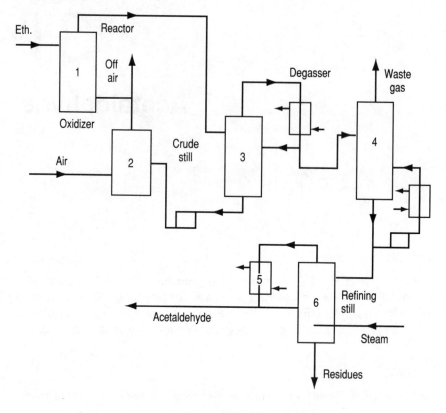

1. Titanium (R50400)
2. Carbon steel
3. - 5. S30403
6. S30403 or copper

Figure 31.1   Acetaldehyde (two-stage).

Ethyl alcohol oxidation

In this process, alcohol vapors are oxidized with air over a silver catalyst at about 480°C (895°F).

Materials

Aluminum alloys

Aluminum is excellent for handling acetaldehyde and related products.

Steel and cast iron

Steel and cast iron are satisfactory if iron contamination is permissible.

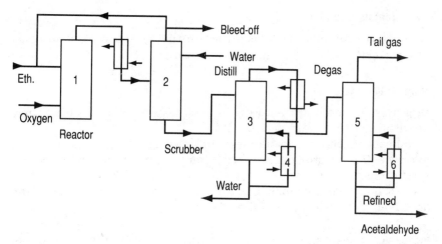

1. - 6.   S30403

Figure 31.2   Acetaldehyde (one-stage).

Stainless steels

Any grade of stainless steel is acceptable for this service. The low-carbon grades are not required, but are often used both because they are "manufacturer's standard" for piping and because they are a precaution against corrosion by acetic acid, a contaminant formed by oxidation of the aldehyde.

Copper and its alloys

Copper and zinc-free alloys have been used for refining acetaldehyde under anaerobic conditions. Discoloration of product by chromophoric corrosion products may result from condensation in the presence of air.

Nickel alloys

Nickel and its alloys are suitably resistant but are generally not required or economical in this service. Alloy 400 (N04400) may be utilized like copper.

Reactive metals

The reactive metals are resistant in this service but have no known application except where waterside chemistry (e.g., seawater cooling) might suggest titanium exchangers. Zirconium is given a slightly lower rating, 2 to 5 mpy (0.5 to 1.2 mm/a) in boiling acetaldehyde.[1]

Precious metals

The precious metals (Ag, Au, Pt) have no current application in this service.

Other metals and alloys

Lead-lined equipment has been used in the older processes to resist acid corrosion, but has no present application.

Nonmetallic materials

Organic. Furane and fluorocarbon plastics are resistant, but others are attacked by solvent action.

Rubber and elastomers likewise suffer solvent attack, but neoprene and some fluorelastomers are satisfactory within specific temperature limitations.[2]

Carbon and impervious graphite equipment is suitable for any acetaldehyde service, as required.

Inorganic. Glass- and stone-ware are satisfactory for any conditions of exposure within normal temperature and pressure limitations.

Pitfalls

In the older processes based on acetylene chemistry, copper-base alloys may form acetylides, a potential ignition source for fire or explosion.

Excessive corrosion of iron or steel may result from oxidation of acetaldehyde to acetic acid.

Storage and Handling

The following materials of construction are commonly used for handling and storage, and their use is thought to constitute good engineering practice:

Tanks	Aluminum, steel, or baked phenolic-coated steel
Tank trucks	Aluminum; phenolic-lined steel
Railroad cars	Aluminum; phenolic-lined steel
Piping	S30403
Valves	Aluminum; CF3M
Pumps	CF3M
Gaskets	Spiral-wound PTFE/stainless steel

References

1. I. Mellan, *Corrosion Resistant Materials Handbook,* Noyes Data Corp., Park Ridge, N.J., 1976.
2. P. A. Schweitzer, *Corrosion Resistance Tables,* 2d ed., Marcel Dekker, New York, 1986.

Chapter

32

Acetate Esters

Introduction

Acetate esters are exemplified by ethyl, isopropyl, butyl, and vinyl acetate. The first three are typically used for solvents in various applications, while the VOAc monomer is a precursor of polyvinyl acetate (a vinyl thermoplastic).

Process

Alkyl acetates

The acetate esters based on saturated hydrocarbons are usually made by reaction between the corresponding alcohol and acid (*esterification*) in the presence of a suitable catalyst (e.g., sulfuric acid). The sulfuric acid may be added directly or premixed with acetic acid or anhydride (acetosulfoacetic acid).

Batch reaction. Traditionally, this is effected in type 316L (S31603) kettles, although continuous operation is sometimes employed. Copper alloys (e.g., silicon bronze, aluminum bronze) have also been used successfully.

In the batch process, the kettle charge is boiled and refluxed until the head temperature indicates (for example) refluxing ethyl acetate at 70°C (158°F). The product can be slowly taken off and fresh alcohol and acid continuously added to the kettle residues.

The major problem is maintaining sufficient catalyst concentrations without inducing corrosion of the stainless steel.

Continuous processes. Two continuous processes have been used. The first is merely the traditional alcohol-acid reaction, while the second is production via the aldehyde (Wacker process).

In the alcohol-acid process, the raw materials are preheated and

169

feed the copper esterifying column. The make comprises a ternary mixture of 20 ester-10 water-70 alcohol which feeds a second separating column. The separation column make is diluted with water, causing separation into two layers, the bottom (water) layer being returned to the lower section of the column and thence as alcohol-water tails to the esterification column. The top layer of crude ester is sent to a drying column, where alcohol and water are taken overhead to be recycled, while the dry ester tails is cooled and stored as final product. Conventional stainless steels (S31603) are preferred to cope with acid corrosion, while the final product is handled in S30400.

Unsaturated esters

The vinyl acetate is produced by direct reaction of acetylene with glacial acetic acid in a continuous process. Type 316L is the standard material, with no particular corrosion problems.

Materials

Aluminum alloys

Aluminum alloys are resistant to acetate esters in themselves, but contamination can lead to some corrosion and haze in the finished product. Free mineral acids and heavy-metal ions have an adverse effect.

Steel and cast iron

Steel and cast iron are generally unacceptable in the process because of acid attack, while iron contamination is unacceptable in the refined product.

Stainless steels

Any of the conventional 18-8 stainless steels are suitable for handling refined acetate esters. Type 316L (S31603) is preferred to cope with traces of acetic acid and to minimize iron contamination.

The high-performance grades (e.g., N08020, S31254, N08367, etc., and their cast analogs) are suitable when traces of sulfuric acid are involved.

Copper and its alloys

Copper and its high-strength, zinc-free alloys, e.g., silicon bronze (C65400) and aluminum bronze (C61400), have been used routinely for reaction and distillation equipment. Corrosion becomes excessive only in the simultaneous presence of acidity and dissolved oxygen, although some discoloration of product may occur.

Nickel alloys

Nickel-base alloys are suitably resistant but find no application because of the availability of less-expensive metals and alloys.

Reactive metals

Titanium would resist the esters themselves but not the combination of organic acids and strong reducing acid catalysts, unless passivated by cupric ion contamination.

The reactive metals (Ti, Zr, Ta) find little or no application in this service.

Precious metals

Although silver has good resistance to acetic acid, the precious metals have no reported application in this service.

Other metals and alloys

There are no reported applications of alternative metals and alloys in acetate ester service. Lead would be adversely affected by traces of acetic acid.

Nonmetallic materials

Organic. Polypropylene and fluorinated plastics are resistant, whereas others suffer solvent attack.

Butyl rubber is useful at ambient temperatures, but other rubbers and elastomers are attacked.

Inorganic. Glass and glass-lined steel are unaffected by any reported conditions in acetate ester service.

Pitfalls

Aluminum may suffer corrosion if the esters contain heavy-metal ions (i.e., by cementation), halide ions, or excess acidity.[1]

Crevice corrosion may occur where type 316L heating coils are rolled into stainless-steel tubesheets, because of local accumulation of sulfuric acid and concentration by hot-wall effects.

A common problem is inadvertent use of nonmolybdenum grades of stainless steel (e.g., S30403, S34700) in lieu of the required S31603. Because mill and warehouse markings and mill analyses are not totally reliable, qualification tests to verify low carbon content and materials identification programs (MIP) to verify the presence of molybdenum by spot test or spectroscopy are recommended for all components.

Storage and Handling

The following materials of construction are commonly used for handling and storage, and their use is thought to constitute good engineering practice:

Tanks	Type 304 or aluminum
Tank trucks	Type 304, aluminum, or baked phenolic-coated steel
Railroad cars	Type 304, aluminum, or baked phenolic-coated steel
Piping	Type 304L or aluminum
Valves	CF3M or aluminum
Pumps	CF3M or aluminum
Gaskets	Spiral-wound PTFE/stainless steel, graphite fiber

References

1. J. A. Lee, *Materials of Construction for Chemical Process Industries*, McGraw-Hill, New York, 1950.

Acetic Acid

Introduction

Acetic acid is a colorless liquid with a sharp vinegary odor, boiling at about 118°C (245°F). A weak acid, it also has powerful solvent action. By far the largest-volume organic acid produced, its place in the organic chemical industry is analogous to that of sulfuric acid in inorganic processes. Acetic acid is used as a starting point in the manufacture of fibers, plastics, agricultural chemicals, pharmaceuticals, and other products.

Process

Historically, acetic acid was produced by fermentation of grain and by destructive distillation of wood. Such older processes have given way to three major routes: (1) acetaldehyde oxidation, (2) liquid-phase hydrocarbon oxidation, and (3) methanol carbonylation.

Acetaldehyde process (Fig. 33.1)

Acetaldehyde in 5 to 15 percent concentration in an acetic acid solution is catalytically oxidized at 50° to 80°C (120° to 175°F) under 0.8 Mpa (100 lb/in^2) pressure. The reaction products are distilled, and gases, esters, and unreacted aldehyde removed. Catalyst carryover and formation of peracids strongly affect materials performance.[1] During distillation, the product is treated with powerful oxidants (e.g., potassium permanganate) to remove oxidizable residual contaminants.

Hydrocarbon oxidation (Fig. 33.2)

A hydrocarbon (e.g., butane) is catalytically oxidized at 185°C (365°F) and 5.2 MPa (750 lb/in^2). Unreacted butane and solvents are removed,

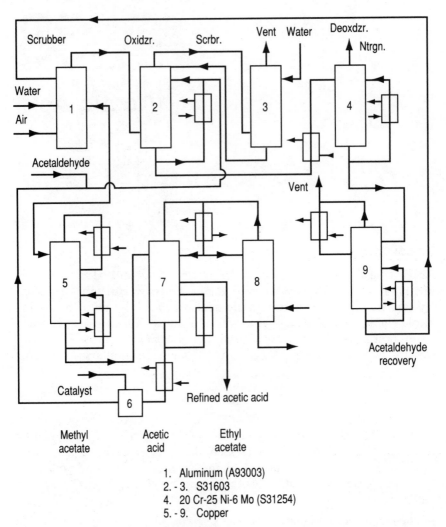

Figure 33.1   Acetic acid: acetaldehyde process.

the acids dehydrated, and formic acid by-product and acetic acid refined by distillation.[2]

Methanol carbonylation (Fig. 33.3)

Methanol and carbon monoxide are reacted at relatively low pressure
in the presence of an iodine-complex catalyst.[3] The reaction liquid
passes through a pressure letdown valve to a flash tank. Vapor from
the flash tank feeds the light ends column, in which catalyst is recovered, light ends going overhead. A side stream feeds the drying col-

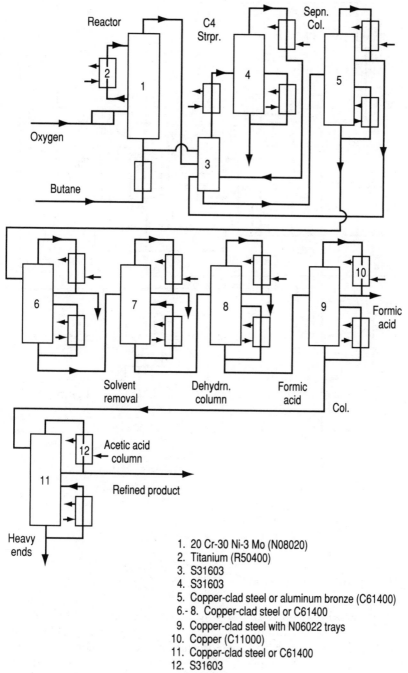

Figure 33.2 Acetic acid: hydrocarbon oxidation.

1. 20 Cr-30 Ni-3 Mo (N08020)
2. Titanium (R50400)
3. S31603
4. S31603
5. Copper-clad steel or aluminum bronze (C61400)
6.- 8. Copper-clad steel or C61400
9. Copper-clad steel with N06022 trays
10. Copper (C11000)
11. Copper-clad steel or C61400
12. S31603

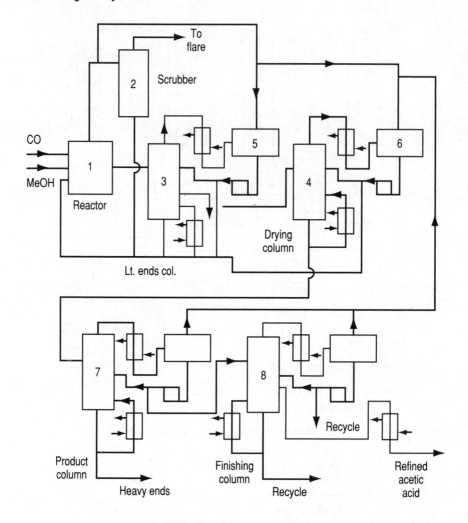

1.   Ni-Mo (N10665)
2.   S31603
3., 4.   N10665
5., 6.   S31603
7., 8.   All S31603

Figure 33.3   Acetic acid: methanol carbonylation.

umn for removal and recycling of water, the tails stream feeding the
product column. Materials of construction are alloy B-2 (N10665) or
zirconium 702 (R60702) in the reactor and flash tank to cope with acetic acid and iodine compounds at 150°C (300°F), while crude columns
are R06702 with S31603 or S31703 overhead equipment. The product
column is S31603.

Materials

Aluminum alloys

Aluminum alloys A91100, A93003, A95052, A95154, and A96061 are used for a variety of equipment handling acetic acid as product.[1] In storage of glacial acid, a haze may develop, and heating coils should be of S31603 construction to prevent hot-wall attack. Corrosion behavior is variable in dilute solutions and adversely affected by the influence of halide or heavy-metal contamination.

Steel and cast iron

Steel and cast iron are not suitable in acetic acid processes or for storage and handling.

Stainless steels

The low-carbon, molybdenum-bearing grade (S31603) is the workhorse material for acetic acid manufacture. Potential problems are intergranular corrosion (unless the low-carbon limitation is rigorously enforced), selective weld corrosion due to molybdenum segregation, and crevice corrosion. The last two phenomena are combated by using an overmatching weld rod (e.g., N08022) for fabrication and for weld overlay of flange surfaces against concentration cell attack.

Copper and its alloys

Although yellow brasses are subject to dezincification, copper and its alloys have excellent resistance in the total absence of oxygen or oxidizing agents. However, corrosion is severe when oxidants are present and even slight attack yields discoloring corrosion products.

Nickel alloys

Of the chromium-free nickel-base alloys, alloy 400 (N04400) has found some applications in anaerobic situations, but it is no better than copper under ideal conditions. The high-molybdenum alloy B-2 (N10665) has found application in the iodine-catalyzed process.

Alloy 600 offers no advantages, but the chromium-molybdenum alloys (e.g., N06022, N10276) are the ideal materials for acetic acid under extreme conditions (e.g., high temperature and/or with chloride contamination).[4]

Reactive metals

Although high-strength titanium grades are subject to SCC, conventional titanium is resistant to the atmospheric boiling points in all concentrations. Anhydrous acid is potentially capable of corroding titanium, as evidenced by electrochemical studies, but this has not occurred in practice.[5] Successful applications may be a result of presence of oxidizing species as contaminants. Reducing contaminants (e.g., 100 ppm Cl^-, formic acid) accelerate corrosion.[6]

Zirconium likewise resists acetic acid and is to be preferred in the presence of iodide contamination.

Precious metals

Silver is very resistant, and heating coils in this material have been used widely in the past. Silver finds less application today, because of cost and the availability of other suitable materials.

Other metals and alloys

Lead is unsuitable in this service, because of the solubility of its acetate salt.

Nonmetallic materials

Organic. Of the thermoplastic materials, all fluorinated plastics are suitable within normal temperature limitations. Chlorinated polyethers and furanes may be used to about 100°C (212°F). Reinforced thermosetting phenolics may be used to at least 120°C (250°F).

Polyethylene drums are suitable for shipping CP acid.

Rubber is a traditional material for dilute acid (less than 5%). Hard rubber and Buna-N may be used in higher concentration at ambient temperatures, while butyl rubber is resistant to about 80°C (175°F).

Carbon and graphite are totally resistant. Impervious graphite heat exchangers, assembled with phenolic cement, have been used for boiling acetic acid/anhydride mixtures.

Inorganic. Glass and ceramic materials are fully resistant, but cement and concrete are rapidly attacked.

Pitfalls

The pitfalls in handling acetic acid relate primarily to contamination effects, i.e., the specific effects of oxidizing or reducing contaminants, halides, etc. on specific alloys susceptible to their effects.

Leakage of cooling waters from condensers can introduce chloride ions into otherwise pure acid, releasing traces of HCl with a variety of consequences (general corrosion, pitting, SCC) in stainless distillation columns. As little as 100 ppm has caused corrosion of S31603. The same degree of chloride ion contamination caused superficial surface hydriding and severe embrittlement of titanium. However, sandblasting removed the hydrided layer and restored ductility.

Storage and Handling

For drum shipment, either solid polyethylene or polyethylene with a steel overpack are used.

The following materials of construction are commonly used for handling and storage of glacial acid (i.e., 99.5% purity), and their use is thought to constitute good engineering practice:

Tanks	Aluminum or S30403
Tank trucks	S31603 or S31600
Railroad cars	S31603 or S31600
Piping	S31603
Valves	CF3M or CF8M
Pumps	CF3M
Gaskets	Spiral-wound PTFE/stainless steel

References

1. I. Mellan, *Corrosion Resistant Materials Handbook,* Noyes Data Corp., Park Ridge, N.J., 1976.
2. "Stainless Steels for Acetic Acid Service," American Iron and Steel Institute, Washington, D.C., April 1977.
3. "A New, Low-Pressure Route to Acetic Acid," *Chemical Engineering,* November 21, 1971.
4. "Corrosion Resistance of Nickel-Containing Alloys in Organic Acids and Related Compounds," International Nickel Co., New York, 1979.
5. A. E. Tsinman et al., *Zaschita Metallov,* vol. 8, 1972, p. 567.
6. *Handbook of Corrosion Data,* ASM International, Metals Park, Ohio, 1989.

Acetic Anhydride

Introduction

The anhydride of acetic acid is a colorless liquid with a choking odor. It is used in organic syntheses, notably in acetylation reactions (e.g., cellulose esters, aspirin).

Process (Fig. 34.1)

One method for manufacture of acetic anhydride $[(CH_3)_2C_2O_3]$ is by coproduction with acetic acid in the dual anhydride process by oxidation of acetaldehyde. The air-oxidation reaction is conducted at about 55°C (130°F) in an acetic acid medium, using a copper/cobalt acetate catalyst, producing acetaldehyde monoperacetate, which decomposes to acetic acid, anhydride, and water.

Low-boiling constituents are stripped in the converter product stripping still (N10276), peracetic acid and diacetyl peroxides are thermally destroyed in a brick-lined kill tank, and acetic acid and acetic anhydride are purified by distillation.

Materials

Aluminum alloys

Aluminum alloys are widely used for the storage and handling of refined product (i.e, tanks, piping, drums, tank cars).[1]

Steel and cast iron

Steel and iron have been used in some older processes in the reaction and distillation/drying steps, with only light to moderate corrosion.[2]

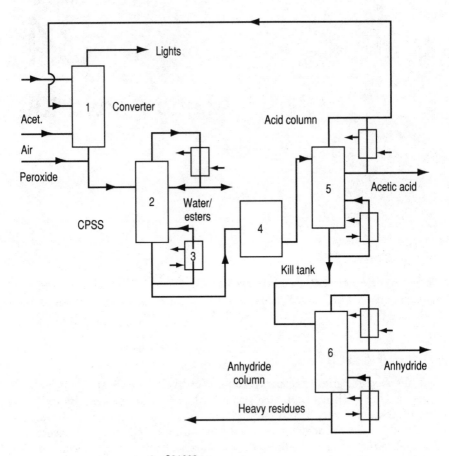

1. S31603
2. Cr-Ni-Mo (N10276)
3. Impervious graphite
4. Brick-lined steel
5. - 6. Copper C11000

Figure 34.1   Acetic anhydride (dual anhydride process).

In modern processes, they are avoided simply because of iron contamination problems.

Silicon-chromium-aluminum alloy steels have been used for a process involving pyrolysis (thermal cracking) of acetic acid vapors.

Stainless steels

Molybdenum-bearing grades of conventional stainless steels (S31603, S31703) are widely used at elevated temperatures.

Copper and its alloys

Copper and its zinc-free alloys, e.g., silicon bronze (C87300), aluminum bronze (C61900), have been used for various pressure vessels, columns, and heat exchangers.

Nickel alloys

The chromium- and molybdenum-free nickel-base alloys offer no advantage over copper alloys in this service.

The Cr-Mo-Ni alloys (N06022, N10276) are ideal for the stills handling boiling acid/anhydride solutions under mildly oxidizing or reducing conditions.

Reactive metals

In the absence of contaminants, titanium, zirconium, and tantalum resist acetic anhydride, but there is little occasion to specify them for this service.

Precious metals

Silver heating coils have been used in some processes, notably in heating acid/anhydride mixtures.

Other metals and alloys

Lead is resistant to acetic anhydride at ambient temperatures, but usually finds no application in this service.

Nonmetallic materials

Organic. Polyethylene is unsuitable, but polypropylene is resistant at ambient temperatures.[3] Otherwise, among the thermoplastic materials, only the fluorinated grades are compatible within their usual temperature limitations. Vinyl esters, bisphenol A-fumarate polyesters, and epoxy- or phenolic-reinforced thermosetting materials are resistant at ambient temperature.

Hard rubber and butyl rubber are resistant to mildly elevated temperatures, 60°C (140°F).

Carbon and graphite are fully resistant (see acetic acid).

Inorganic. Glassware and ceramics are resistant, but concrete and cement are attacked by acid formed by hydrolysis of the anhydride.

Pitfalls

Unforeseen corrosion is likely only in the event of contamination by halides or by oxidizing or reducing species.

Storage and Handling

Acetic anhydride is hazardous, causing painful burns and blisters. Protective clothing, goggles, and masks should be used in handling.

Use of the following materials of construction is common for handling and storage, and is thought to constitute good engineering practice:

Tanks	Aluminum
Tank trucks	Aluminum
Railroad cars	Aluminum
Piping	Aluminum
Valves	A03560 or CF8M
Pumps	CF3M
Gaskets	Butyl rubber, spiral-wound PTFE/stainless steel, flexible graphite

References

1. *Handbook of Corrosion Data*, ASM International, Metals Park, Ohio, 1989.
2. J. A. Lee, *Materials of Construction for Chemical Process Industries*, McGraw-Hill, New York, 1950.
3. P. A. Schweitzer, *Corrosion Resistant Materials*, 2d ed., Marcel Dekker, New York, 1986.

Chapter

35

Acetone

Introduction

Acetone (dimethyl ketone) is a colorless, flammable liquid with a pungent odor. Miscible with water as well as other organic solvents (ethers, alcohols, and esters), it is used both as a solvent (e.g., esters, plastics, acetylene) and as a reactant in organic syntheses (e.g., ketene, mesityl oxide, isophorone).

Process

A fermentation process from cornstarch and molasses became obsolete with the development of organic syntheses (e.g., cumene hydroperoxide to phenol, isopropyl alcohol dehydrogenation).

Cumene hydroperoxide process (Fig. 35.1)

Benzene is alkylated to cumene, oxidized to the hydroperoxide and cleaved to produce phenol and acetone.

Usually, there is a series of oxidizers operating at about 80° to 130°C (175° to 265°F) under pressure, with an alkaline promoter (e.g., sodium hydroxide). The hydroperoxide is concentrated, then cleaved under acid conditions and water-washed to remove salts. Light ends are distilled off, and the acetone refined in a final column.

Isopropanol dehydrogenation (Fig. 35.2)

Isopropanol is catalytically dehydrogenated at about 380°C (715°F) in a hydrogen atmosphere in a tubular reactor followed by a fixed-bed reactor. The reaction product contains unconverted alcohol and miscellaneous light ends (hydrocarbons, aldehydes, CO, etc.) as well as acetone. Noncondensables are scrubbed with water and recycled. The liquids are separated by distillation, alcohol being recycled and acetone purified after caustic treatment to remove aldehydes.

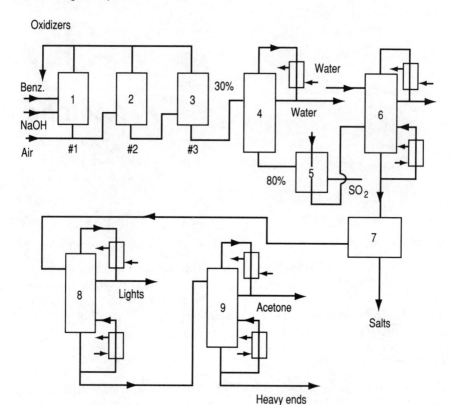

1. - 4. S30403
5. - 7. S31603
8. - 9. Carbon steel with 17 Cr trays (S43000)

Figure 35.1   Acetone via cumene.

Materials

Aluminum alloys

Aluminum and its alloys are resistant to acetone and its water solutions, in the absence of chlorides, heavy-metal salts, etc., as water-borne contaminants.

Steel and cast iron

Acetone is noncorrosive to iron and steel, but water solutions become corrosive with dissolved oxygen and salt contamination.

Stainless steels

All grades of stainless steels are fully resistant to acetone. Localized corrosion may occur in water solutions, occasioned by chlorides

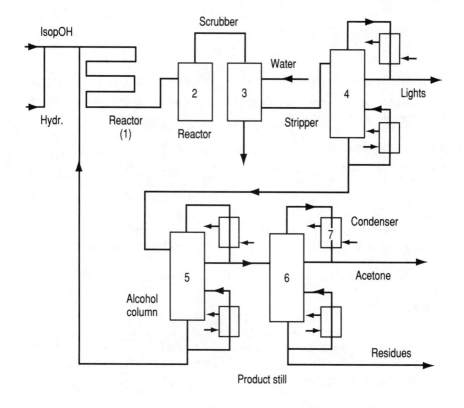

1. - 2. 2 1/2 Cr-1 Mo (K21390)
3. - 6. Carbon steel with 17 Cr trays (S43000)
7. S30403

Figure 35.2 Acetone via isopropanol.

and other contaminants contributing to underdeposit oxygen concentration-cell effects.

Copper and its alloys

All copper alloys are resistant to acetone.

Nickel alloys

All nickel alloys are fully resistant.

Reactive metals

The reactive metals are immune to attack, but find no applications in this service.

Precious metals

Silver, gold, and platinum are inert in this service.

Other metals and alloys

Lead is practically resistant [less than 0.5 mm/a (20 mpy)] in acetone, but finds no application.

Nonmetallic materials

Organic. Polyethylene is resistant at ambient temperature and polypropylene to about 80°C (175°F). Fluorocarbons *above* PVDF are resistant. The reinforced thermosetting materials are generally unsuitable.

Butyl and silicone elastomers are resistant at ambient temperature. Otherwise, rubber and elastomers are generally unsuitable.[1]

Inorganic. Nonmetallic materials are unaffected by acetone.

Pitfalls

There are no reported pitfalls with acetone, except contamination effects in water solutions. Stainless steels have failed by the combined effect of iron oxide films and chloride retention in recovery of acetone by distillation from water solutions.

Storage and Handling

Acetone is flammable, with a moderate explosion hazard.

Use of the following materials of construction is common for handling and storage and is thought to constitute good engineering practice:

Tanks	Steel
Tank trucks	Steel
Railroad cars	Steel
Piping	Steel
Valves	Steel or CI/S41000 trim
Pumps	Steel or CI
Gaskets	Butyl rubber, spiral-wound PTFE/stainless steel

References

1. P. A. Schweitzer, *Corrosion Resistance Tables*, 2d ed., Marcel Dekker, New York, 1986.

Acrolein

Introduction

Acrolein (2-propenal), the simplest unsaturated aldehyde, is an important building block in organic syntheses. A colorless, volatile liquid, it has strong lachrymatory effects. It must be inhibited (e.g., with hydroquinone) to prevent exothermic polymerization under the influence of heat or light. Trace amounts (e.g., 6 to 10 ppm) are used for microbiological control in water systems and jet fuels.

Process (Fig. 36.1)

The original process of acetaldehyde/formaldehyde condensation has been replaced by catalytic oxidation of propylene.

Propylene and air are preheated and reacted at about 300° to 400°C (570° to 750°F) under 200 to 300 kPa (30 to 45 lb/in² pressure. Reaction products are cooled and quenched, acrylic acid being removed. Crude acrolein is stripped of noncondensables and light ends and refined by final distillation.

Materials

Aluminum alloys

Aluminum is inherently resistant to acrolein. Aluminum process equipment was widely used in the formaldehyde/acetaldehyde process.[1]

Steel and cast iron

Steel and cast iron are resistant to acrolein.

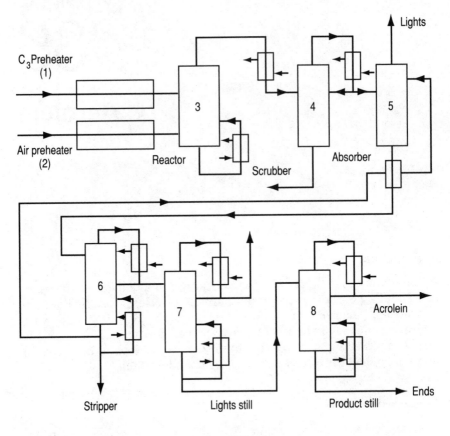

1. - 2. 5 Cr Steel (S50100)
3. - 5. S30403
6. - 7. S31603
8. Carbon steel with S30403 trays

Figure 36.1   Acrolein (propane oxidation).

Stainless steels

Austenitic stainless steels (e.g., S30403) have been used to resist high-temperature oxidation.

Copper and its alloys

Copper alloys are resistant to acrolein by itself. Acid contamination in the presence of oxidants would cause corrosion and discoloration.

Nickel alloys

There are no known applications for nickel alloys in this service.

Reactive metals

Reactive metals are unaffected by acrolein but find no application.

Precious metals

Silver, gold, and platinum are fully resistant.

Other metals and alloys

Lead reportedly is practically resistant [less than 0.5 mm/a (20 mpy)] in a 10% water solution.[1]

Nonmetallic materials

Organic. No data were located. By analogy, the data relevant to acetaldehyde (see Chap. 31) should be applicable.

Inorganic. Inorganic materials are unaffected by acrolein. The alkalinity in cement or concrete would cause polymerization.

Pitfalls

Acrolein can polymerize, with a strong exothermic reaction and potentially explosive force, if inadvertently contaminated (e.g., by residues of soda ash in a steel tank car).

Storage and Handling

Acrolein must be inhibited against polymerization and protected from heat, light, and contamination by acids, alkalies, etc. The pH is adjusted to 5 to 6 with acetic acid to prevent aldol condensation. Storage systems must be dedicated, and no hoses or lines should be used that might contain residual chemicals from other services. Acrolein is highly toxic to most forms of life, and air or water contamination must be avoided. (Dilute alkalies and sodium bisulfite treatment can render small spills biodegradable.)

Use of the following materials of construction is common for handling and storage and is thought to constitute good engineering practice:

Tanks	Steel
Tank trucks	Steel
Railroad cars	Steel (baked phenolic permissible)
Piping	Steel

Valves	Steel or cast iron/S41000 trim
Pumps	Steel
Gaskets	PTFE, silicone rubber

References

1. I. Mellan, *Corrosion Resistant Materials Handbook,* Noyes Data Corp., Park Ridge, N.J., 1976.

Acrylic Acid and Esters

Introduction

Acrylic (propenoic) acid is the basic unsaturated acid and is therefore capable of polymerization. Both the acid and its esters are used to prepare emulsion and solution polymers which find application in paints and coatings, polishes, and adhesives.

Process

Acrylic acid (Fig. 37.1)

Processes based on acetylene are being replaced by propylene oxidation in a two-step process. The first step involves production of acrolein (see Chap. 36), which is oxidized in the second step to acrylic acid. When acrolein is not desired as a specific product, there is no refining until the acrylic acid separation, the reaction products being absorbed in 20% aqueous acrylic acid.

Reactors are fixed-bed shell-and-tube type, the tubes about 12 to 15 ft (3.65 to 4.57 m) long and 1 in (25.4 mm) in diameter, packed with catalyst, and cooled with molten salt. Vaporized propylene and an air-steam mixture compose the feed. Oxidation to acrolein proceeds in the no. 1 reactor and is continued to acrylic acid in the no. 2 reactor, waste gas, including carbon monoxide and dioxide, being removed. The dilute aqueous acrylic acid (20 to 30%) is cooled and sent to the separation step.

The acrylic acid is extracted with a suitable solvent (e.g., esters, ketones) and the extract distilled under vacuum (to maintain a low temperature) in a solvent-recovery column. Tails from solvent recovery

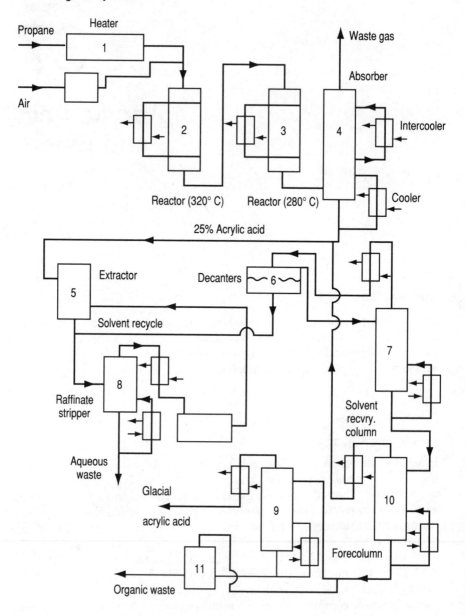

Propane
Heater

Air

Reactor (320° C)   Reactor (280° C)

Waste gas

Absorber

Intercooler

Cooler

25% Acrylic acid

Extractor

Decanters

Solvent recycle

Raffinate
stripper

Aqueous
waste

Glacial

acrylic acid

Solvent
recvry.
column

Forecolumn

Organic waste

1. 5 Cr steel (S50100)
2. - 7. S31603
8. - 11. S30403

Figure 37.1 Acrylic acid (propane oxidation).

feed the foreruns column where water, acetic acid, and solvent are taken overhead. Glacial acrylic acid of about 98% purity is distilled in the acrylic acid column.

Acrylic esters (Fig. 37.2)

Acrylic acid, alcohol, and a catalyst (e.g., sulfuric acid) are fed to a glass-lined reactor with an external calandria and a distillation column. The organic distillate of ester/alcohol/water is sent to the wash column, whose tails is fed to a dehydration still. In the dehydration still, water is taken overhead, while the dry bottoms feed the product column to produce the refined ester (e.g., ethyl acrylate).

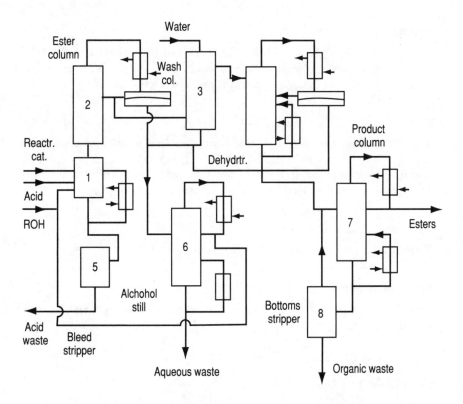

1. S31603
2. Brick-lined steel
3. - 6. S31603
7. - 8. S30403

Figure 37.2   Acrylic esters.

Materials

Aluminum alloys

Aluminum and its alloys are resistant to acrylic acid and to the esters, but not to 50% acid solutions in water, alcohols, or toluene.[1,2]

Steel and cast iron

Iron and steel are incompatible with acrylic acid. High-silicon iron (F47003) is resistant but finds no application in this service.

Stainless steels

The molybdenum-bearing grade S31603 is a standard material for acrylic acid, analogous to acetic acid service. S30400 is suitable for shipment and storage.

Copper and its alloys

Except for the high-zinc yellow brasses, copper alloys resist acrylic acid in the absence of oxidizing species. They are not used for refined products because of chromophoric corrosion products.

Nickel alloys

The chromium-free alloys resist acrylic acid in the absence of oxidizing species. However, nickel (N02200) is deemed unsatisfactory for shipping containers because of chromophoric compounds. The Cr-Ni-Mo grades (N06625, N06022, N10276) are fully resistant.

Reactive metals

Titanium, zirconium, and tantalum are resistant.

Precious metals

Silver is resistant, but there is no occasion to use the precious metals in this service.

Other metals and alloys

Lead and zinc are unsuitable in acrylic acid.

Nonmetallic materials

Organic. Polyethylene and polypropylene resist glacial acid and aqueous solutions. Fluorinated plastics are resistant. Acrylic acid is expected to behave toward other organic materials and elastomers analogously to acetic acid, with the added problem of possible catalytic polymerization caused by contained pigments and additives.

Carbon and graphite are resistant, but are not employed because of their susceptibility to breakage when polymeric deposits must be removed.

Inorganic. Glass and ceramics are compatible with the acid and esters. Cement and concrete should not be exposed to these materials.

Pitfalls

The possibility of runaway polymerization must always be considered.

S31603-clad steel vessels have suffered SCC when the acid became chloride-contaminated.

Storage and Handling

Acrylic acid and esters must be stabilized to prevent polymerization [e.g., with the monomethyl ether of hydroquinone (MEHQ)] and stored under air at room temperature (acrylic acid freezes at 13°C).

Use of the following materials of construction is common for handling and storage and is thought to constitute good engineering practice:

	Acrylic acid	Acrylic esters
Tanks	S30403	Steel, stainless steel, aluminum
Tank trucks	S30403	Steel, stainless steel, aluminum
Railroad cars	S30403	Steel, stainless steel, aluminum
Piping	S30403	Steel, stainless steel, aluminum
Valves	CF8M	Steel/S41000
Pumps	CF8M	CF8M
Gaskets	PTFE/SS	PTFE/stainless steel

References

1. *Corrosion Resistant Materials Handbook*, 3d ed., Noyes Data Corp., Park Ridge, N.J., 1976.
2. Private communication.

Chapter

38

Alkanolamines

Introduction

Monoalkanolamines have the generic formula NH_2ROH, exemplified by monoethanolamine (MEA, $NH_2C_2H_4OH$). The polyalkanolamines are such products as diethanolamine (DEA) and diglycolamine (DGA). Alkaline water-soluble solvents, they are widely used in acid-gas scrubbing systems and in organic syntheses.

Process (Fig. 38.1)

Alkanolamines are produced by reaction of aqueous ammonia and an organic oxide (e.g., ethylene oxide or its homologs). Excess ammonia is flashed off and recycled; the resulting aqueous mixture of mono-, di- and triethanolamines is separated by distillation. The proportion of the products can be varied by changing temperatures, pressures, and ratios of ammonia to oxide (always with an excess of ammonia) or by use of diluent gases.

Materials

Aluminum alloys

Aluminum is inherently resistant to the concentrated solutions at ambient temperatures. MEA freezes at 10.5°C (50°F), and aluminum is not suitable for heating coils because of hot-wall attack. Below about 60% concentration in water, the solution is too alkaline for aluminum. However, solutions of 20% MEA in 72.5% glycol/7.5% water are handled in aluminum in the simultaneous drying and sweetening of hydrocarbon gases.

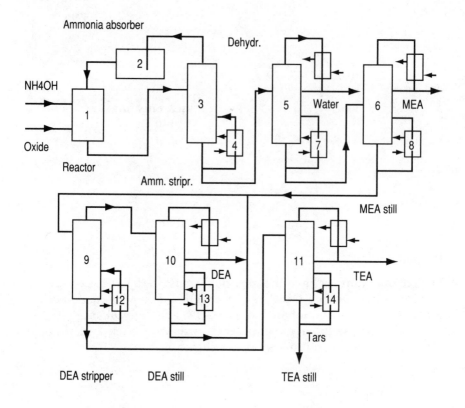

Figure 38.1 Monoethanolamine.

1. S30400
2. Carbon steel
3. S30400 column/ S31600 trays
4. S31600
5.–6. Carbon steel column/ S43000 trays

7.–8. S30400
9.–11. Carbon steel/ S3000 trays
12.–13. S31600
14. S30400

Steel and cast iron

Steel and cast iron are resistant to MEA alone and to aqueous solutions in the absence of excess acid gases. Iron contamination and discoloration is a possible problem.

Stainless steels

Any grade of stainless steel is resistant to the alkanolamines themselves. At elevated temperatures, or in the presence of acid gas contaminants, the molybdenum-bearing grades such as S31600 are preferred. Intergranular corrosion is not a problem in the MEA itself, so

S31603 is not required. The higher alloys like N08020 are *not* better than S31600, because of nickel/amine complexes.

Copper and its alloys

Copper alloys can be used in alkanolamines only under rigorously nonoxidizing conditions, forming the royal-blue cupric/amine complexes otherwise with attendant high rates of attack. They find little or no practical application for this reason.

Nickel alloys

The chromium-free nickel alloys form metal/amine complexes in the simultaneous presence of oxidizing agents. The chromium-bearing alloys are more resistant but less so than the appropriate stainless steels.

Reactive metals

The reactive metals find no application in the manufacture of alkanolamines. Tantalum would be attacked by aqueous solutions.

Precious metals

Silver should not be exposed to ammonia or its derivatives, because of the possible formation of explosive compounds (i.e., azides).

Other metals and alloys

Lead, zinc, and tin are nonresistant to aqueous solutions because of their alkaline nature, and zinc and tin coatings are incompatible with the refined products. Lead is moderately resistant to cold alkanolamines but finds no useful application in lieu of carbon steel.

Nonmetallic materials

Organic. MEA has a powerful solvent action against most conventional organic materials. Of the thermoplastic materials, polypropylene is suitable to about 80°C (175°F), above which temperature only the fluoroplastics *above* PVDF are resistant.[1]

Of the elastomers, only Viton and Buna N are useful to 80°C (175°F).

Inorganic. Glass and ceramics are resistant to the product by itself, but suffer alkaline attack in aqueous solutions.

Pitfalls

Aluminum is very satisfactory for handling and storage of undiluted MEA, for example. However, it cannot safely be used for *heating coils,* because of sudden corrosion after an initiation period, a result of hot-wall effects. Inadvertent dilution will cause serious corrosion.

Carbamate corrosion may occur in ammonia recovery systems, because of CO_2 contamination. See Chap. 64.

Warning: The materials information in this chapter should *not* be applied to alkanolamine solutions in acid-gas scrubbing systems. Corrosion in those systems is governed by acid gases (e.g., sulfide phenomena), mole ratios, thermal degradation products, and other process-related effects.

Storage and Handling

Use of the following materials of construction is common for handling and storage and is thought to constitute good engineering practice:

Tanks	Aluminum
Tank trucks	Aluminum (with type 316L heating coils for MEA or whenever heating is required)
Railroad cars	Aluminum (S31600 coils)
Piping	S30403
Valves	CF8M or A035600
Pumps	CF8M
Gaskets	Buna N, spiral-wound PTFE/stainless steel, flexible graphite

References

1. P. A. Schweitzer, *Corrosion Resistance Tables,* 2d ed., Marcel Dekker, New York, 1986.

39

Alkylamines

Introduction

Alkylamines (e.g., methylamine, ethylamine) are commonly produced by catalytic reaction of anhydrous ammonia with the corresponding alcohol (e.g., the Leonard process). The alkylamines are used in explosives, insecticides, and surfactants, and as precursors in specific processes.

Clear liquids with a fishy and ammoniacal odor, they are powerful solvents for plastics, coatings, and elastomers, with a strong complexing power for some metal ions (e.g., copper, nickel, iron).

Process (Fig. 39.1)

Anhydrous ammonia and alcohol vapors are passed through an S31603 catalytic converter, where

$$ROH + NH_3 \rightarrow RNH_2 + H_2O$$

and the crude vapors enter an ammonia stripping still. The still has traditionally been of carbon steel construction with S410 valve trays. However, intermittent corrosion problems (see "Pitfalls" below) have led to a preference for S30403-clad steel with S31603 trays.

The ammonia column tails, comprising crude amines, ammonium salts, and water, feed four steel product columns with S30403 trays, where the several products (various alkylamines) are refined and accumulated in product storage tanks. Water and waterborne salts feed the vent column, which is sparged with high-pressure steam. Vent column tails is recirculated to the vaporizer.

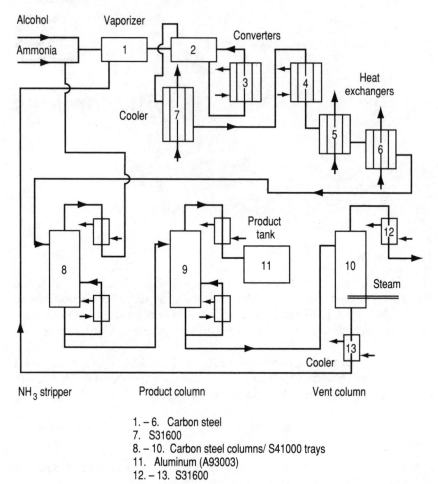

Alcohol
Ammonia
Vaporizer

Converters

Cooler

Heat
exchangers

Product
tank

Steam

Cooler

NH$_3$ stripper   Product column   Vent column

1. – 6. Carbon steel
7. S31600
8. – 10. Carbon steel columns/ S41000 trays
11. Aluminum (A93003)
12. – 13. S31600

Figure 39.1   Alkylamines (Leonard process).

Materials

Aluminum alloys

Aluminum and its alloys are resistant to anhydrous alkylamines. They will be attacked in aqueous solutions (e.g., at 72% concentration).

Cast iron and steel

Cast iron and steel are resistant to anhydrous amines, but some iron contamination will occur at moderately elevated temperatures [e.g., greater than 40°C (104°F)].

Stainless steels

Types 304 (S30400) and 304L (S30403) are resistant to alkylamines to about 80°C (175°F), above which temperatures the molybdenum-bearing S31603 is required.

Copper alloys

Copper and its alloys are not practical in amine services, since all oxygen and oxidants must be excluded.

Nickel alloys

Nickel alloys find no application in alkylamine services.

Reactive metals

Titanium and zirconium resist aqueous amine solutions but are neither required nor economical in alkylamines.

Noble metals

Gold and platinum resist this type of service but are not utilized. *Silver* should not be exposed to ammonia derivatives because of the danger of silver azide formation.

Other metals

Lead and zinc are not satisfactory in alkylamines, but tin coatings are acceptable for both refined ethylamine and the 70 to 72% aqueous solution.

Nonmetals

Organic. Amines are powerful solvents and only fluoroplastics above PVDF would be suitable in such services.

Impervious graphite is suitable for heat exchangers provided a resistant cement (not phenolic-based) is used in assembly.

Inorganic. Glass and ceramicware are resistant but are not required in this process.

Pitfalls

The major pitfall in this process is that there may be an accumulation of carbon dioxide in the ammonia stripping still, which leads to carbamate corrosion. Traces of CO_2 may be formed in the converters (e.g., because of catalyst deterioration or overheating) and will accumulate as ammonium carbamate in the NH_3 stripping still, unable to go overhead with the ammonia and unable to exit in the tails as ammonium carbonate because of thermal decomposition. Severe localized thinning of the column walls in the feed section, and localized failure of trays and tray fittings can occur. This corrosion zone can then progress upward or downward as the column distillation characteristics change. Periodic nondestructive inspection of the feed-section column walls is imperative. The critical CO_2 contamination level is about 300 ppm in ammonia containing about the same amount of water vapor, when temperatures are 120°C (250°F) at autogenous pressures. See Chap. 60.

Another pitfall is that alkaline boiler-treating compounds from high-pressure steam to the vent stripper may reach the vaporizer and cause caustic SCC.

Storage and Handling

Use of the following materials of construction is common for handling and storage and is thought to constitute good engineering practice:

Tanks	Aluminum, type 304
Tank trucks	Aluminum, type 304
Railroad cars	Aluminum, type 304
Piping	Type 304
Valves	CF8M
Pumps	CF8M
Gaskets	Spiral-wound PTFE/stainless steel

40

Alkylene Polyamines

Introduction

The ethylene amine family consists of ethylene diamine (EDA) itself and its higher homologs, such as diethylene triamine (DTA), tetraethylene pentamine (TPA), etc. These are slightly viscous liquids with a strong ammoniacal odor. They are strong bases with distinct chelating and solvent properties. The alkylene polyamines are primarily chemical intermediates in the manufacture of fungicides, chelating agents, resins, and surfactants.

Process (Fig. 40.1)

Although EDA can be produced by catalytic amination of MEA, the primary process is reaction of aqueous ammonia with ethylene dichloride (EDC). This reaction yields the amine hydrochloride, which must be neutralized with caustic to spring the free amines. The family of higher homologs is also produced as by-products.

A multistage pump feeds ammonium hydroxide and EDC to a type 316L tubular reactor at about 150°C (302°F) and 6.9 Mpa (1000 lb/in^2). Unreacted ammonia is flashed from the resulting aqueous amine hydrochlorides, which are neutralized in a titanium section of the 316L ammonia column. A water column strips off water and salt is centrifuged from the amine mixture. The crude amine mixture is separated by distillation in successive steps (i.e., in EDA and DTA columns under vacuum). The polyamines, including Piperazine (diethylene diamine) may be further separated as required.

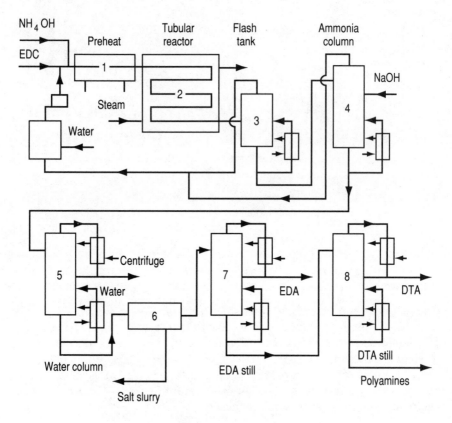

1.   Carbon steel
2.   S31603
3. – 5.   Cr-Ni-6 Mo (S31254)
6.   Cr-Ni-Mo (N10276)
7. – 8.   S31603

Figure 40.1   Alkyleneamines.

Materials

Aluminum alloys

Aluminum is suitable for EDA and its homologs at ambient temperatures. The 70% aqueous solution is corrosive.

Steel and cast iron

Steel and cast iron are acceptable in EDA itself, except for iron contamination, which may be unacceptable to some end uses. They are unacceptable in the 70% solution, which is more prone to iron contamination.

Stainless steels

The austenitic stainless steels are suitable for storage. The molybdenum-bearing grades (e.g., S31603, S31703) are required in the process. Higher nickel alloys (e.g., N08020) offer no advantage because of complex corrosion products.

Copper and its alloys

The copper alloys are generally unacceptable unless air and oxidants can be totally excluded. Copper equipment has been used for dehydration of EDA with strong caustic solutions.

Nickel alloys

Alloy 400 has been used with moderate success for salt-contaminated process liquors. It is attacked by ingress of air or oxidizing agents. Alloy C276 (N10276) has been used for salt centrifuges in this service. Otherwise, nickel alloys are not usually employed, for reasons previously stated.

Reactive metals

Titanium, zirconium, and tantalum are not normally used. Specific column *sections* have been fabricated in titanium and zirconium to resist localized carbamate corrosion.

Precious metals

Silver must be excluded from this service, while gold and platinum find no application.

Other metals and alloys

Zinc is nonresistant, but tinned steel is suitable for small containers in both pure and 70% solutions.

Nonmetallic materials

Organic. Only FEP and PTFE fluorocarbons are suitable. PVDF reportedly suffered environmental cracking at ambient temperature.[1] Reinforced thermosets are nonresistant.

Buna N is reportedly resistant at ambient temperatures.

Carbon and graphite are resistant, but the phenolic and epoxy cements and binders are not.

Inorganic. Glass and ceramics are resistant to the amines themselves but are attacked by aqueous solutions at moderately elevated temperatures.

Pitfalls

The chloride contamination inherent in the process is highly conducive to pitting and stress-corrosion cracking.

Localized corrosion in sections of the ammonia and water columns may occur because of carbamate attack if carbon dioxide gains access (e.g., via the ammonium hydroxide or the caustic neutralization).

Storage and Handling

Amine vapors are an eye irritant and painful to nose, throat, and respiratory system. Also, sensitive people may develop dermatitis or an asthmatic response. Goggles and protective clothing should always be worn.

Use of the following materials of construction is common for handling and storage and is thought to constitute good engineering practice:

Tanks	Aluminum, S30400
Tank trucks	Aluminum, S30400
Railroad cars	Aluminum, S30400
Piping	S30403
Valves	CF8M
Pumps	CF8M
Gaskets	Buna N, spiral-wound PTFE/stainless steel, braided graphite

References

1. I. Mellan, *Corrosion Resistant Materials Handbook*, 3d ed., Noyes Data Corp., Park Ridge, N.J., 1976.

Caprolactam

Introduction

A white, hygroscopic, crystalline solid, caprolactam is a widely used chemical intermediate, primarily in the manufacture of nylon-6 fibers and polyamide resins and plastics.

Process

Caprolactam is usually produced either from benzene or toluene.

Benzene processes

Benzene may either be alkylated to phenol (which is catalytically hydrogenated to cyclohexanone) or hydrogenated to cyclohexane.

Cyclohexanone route (Fig. 41.1). Ammonium nitrite solution is reduced with sulfur dioxide and hydrolyzed. The acidic hydroxyl-amine sulfate is simultaneously neutralized with ammonia and reacted with cyclohexanone to form the cyclohexanone oxime. This yields a molten oxime layer and a saturated ammonium sulfate solution. The oxime undergoes molecular (Beckman) rearrangement in the presence of oleum to yield caprolactam.

Cyclohexane route. Cyclohexane may be photonitrosated to cyclohexanone oxime hydrochloride via nitrosyl chloride and HCl. The molecular rearrangement in oleum produces caprolactam and allows HCl to be recycled for nitrosyl chloride production.

Toluene process

Toluene is catalytically oxidized with air to benzoic acid. The benzoic acid is catalytically hydrogenated to cyclohexanecarboxylic acid,

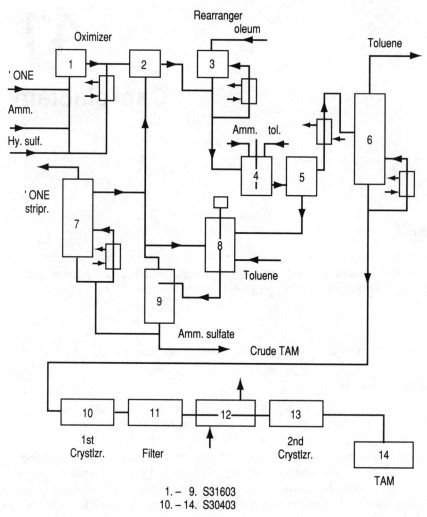

Figure 41.1   Caprolactam (cyclohexanone process).

which is reacted with NO_x in oleum to form crude caprolactam. The reaction mass may be neutralized with ammonia, producing ammonium sulfate. Alternatively, the caprolactam may be diluted with water and extracted with alkylphenol solvent.

Materials

Aluminum alloys

Laboratory tests indicate that caprolactam is discolored by aluminum.[1]

Steel and cast iron

Steel and cast iron are unsuitable for caprolactam service.

Stainless steels

Austenitic stainless steel (S30403) is used to resist corrosion in the caprolactam purification steps and to prevent product contamination. In the crude lactam, S31603 is used for the oximation and rearrangement steps to resist hot ammonium sulfate, although the final dewatered ammonium sulfate by-product is stored in S30400.

Copper and its alloys

Copper and copper alloys find no application in this service.

Nickel alloys

Chromium-free nickel alloys are not compatible with caprolactam. The nickel-chromium-molybdenum grades are not needed in any part of the process.

Reactive metals

Titanium finds application in the hydrolysis of the hydroxylamine disulfonate to produce hydroxylamine.

Precious metals

Silver, gold, and platinum have no applications in this process.

Other metals and alloys

Zinc-pigmented inorganic coatings are alleged to be resistant to caprolactam.[1]

Nonmetallic materials

Organic. In the absence of specific information, only fluorinated plastics (e.g., FEP, PTFE, PFA) seem reliable for this service.

Inorganic. Glass and ceramics would be suitable for caprolactam, but no usage is reported.

Pitfalls

There are no reported pitfalls relating to materials of construction in this process. The handling of ammonium carbonate suggests a possibility of carbamate formation, but the ammonium nitrite reaction assures oxidizing conditions sufficient for stainless steel resistance.

Storage and Handling

Flaked caprolactam is shipped in paper bags with a polyethylene liner. Bulk shipments are stored at about 75°C (160°F) under nitrogen (less than 5 ppm oxygen).

Use of the following materials of construction is common for handling and storage and is thought to constitute good engineering practice:

Tanks	Aluminum or S30400
Tank trucks	Aluminum or S30400
Railroad cars	Aluminum or S30400
Piping	S30403
Valves	CF8M
Pumps	CF8M
Gaskets	Spiral-wound PTFE/stainless steel

References

1. I. Mellan, *Corrosion Resistant Materials Handbook,* 3d ed., Noyes Data Corp., Park Ridge, N.J., 1976.

Chapter

42

Carbon
Tetrachloride

Introduction

Carbon tetrachloride (tetrachloromethane) is a heavy colorless liquid with a characteristic but nonirritating odor. One of the first organic chemicals produced on a large scale, it finds application as a solvent, process intermediate (e.g., refrigerant manufacture), and fumigant. Because it is highly toxic, its prior applications in metal degreasing and dry cleaning have been eliminated. Intermediate halocarbons such as methyl chloride, methylene chloride, and chloroform may be produced as precursors.

Process

The original process of chlorination of carbon disulfide has been largely replaced by methane or methyl chloride chlorination.

Methane is catalytically chlorinated at elevated temperatures (e.g., 300°C) in a nickel (N02200) vessel, producing a series of chlorinated hydrocarbon products (Fig 42.1). Unwanted coproducts are suppressed by appropriate recycle streams. Acid-brick-lined and carbon or graphite vessels are used to withstand the wet acid conditions. Copper may be used for the caustic treatment of crude product, while alloy G30 (N06030) is used for the drying operation with 98% sulfuric acid. Because of the stepwise chlorination, Ni-Cr-Mo alloys (e.g., N10276) and copper alloys are employed.

In another process, methyl chloride is produced by hydrochlorination of methanol, the methyl chloride being subsequently chlorinated (Fig. 42.2). Nickel-base alloys are used to chlorinate the methyl chloride, anhydrous HCl being recovered in a PTFE column. Dry

215

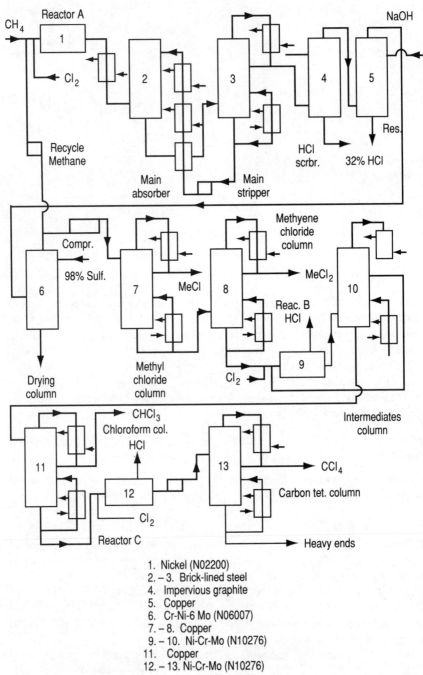

Figure 42.1   Methane chlorination.

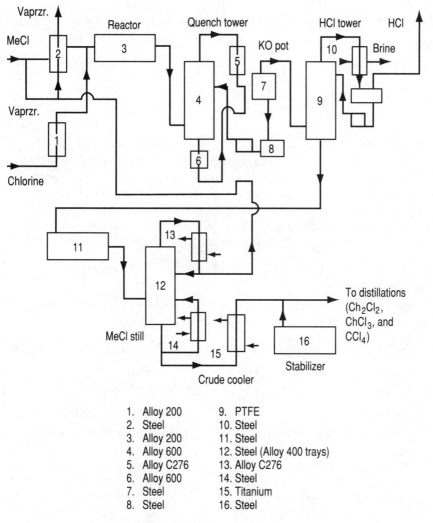

Figure 42.2   Chlorocarbons; Methyl Chloride Chlorination.

chloromethane products are separated in carbon steel columns, although alloy 400 internals may be used.

There are a number of possible variations on these processes and carbon tetrachloride may also be produced by pyrolysis or "chlorinolysis" of longer chain hydrocarbons.

The crude carbon tetrachloride is neutralized, dried, and distilled at a high reflux ratio. Small amounts of stabilizing agents (e.g., alkylcyanamides, amines, fatty acid derivatives) may be added to minimize reactivity with specific alloy systems.

A good technical grade of CCl_4 should contain no more than 1 ppm acidity as HCl, 20 ppm bromine, 200 ppm water, and 150 ppm chloroform ($CHCl_3$).

Materials

Aluminum alloys

Aluminum and its alloys are nominally resistant to carbon tetrachloride up to 60°C (140°F).[1] However, they are attacked at elevated temperatures or when the oxide film is disturbed. Powdered aluminum can react explosively with carbon tetrachloride.[2] In practice, the preferred alloys are those containing magnesium (e.g., A95052, A95154) and temperatures should not exceed 50°C (120°F). Extreme caution should be used in exposing aluminum to any chlorinated solvents.[3]

Steel and cast iron

Anhydrous carbon tetrachloride is noncorrosive. However, even small amounts of water render iron and steel unsuitable because of hydrolysis and acid attack. At only slightly elevated temperatures, *rust* can catalyze decomposition, releasing phosgene ($COCl_2$) as well as HCl. High-silicon iron (F47003) is fully resistant.

Stainless steels

The stainless steels are unaffected by pure carbon tetrachloride. In practice, although both S30400 and S31600 show no general corrosion, pitting, crevice corrosion, and/or SCC due to traces of HCl may ensue in the presence of moisture. The molybdenum-bearing grades would be preferred as somewhat more resistant.[4]

Copper and its alloys

Except for yellow brasses, copper and its alloys are considered resistant to carbon tetrachloride even in the presence of moisture. As always, the presence of air or oxidants and any accumulation of Cu^{++} ions could give unexpectedly high rates.

Nickel alloys

N04400 would have substantially the same characteristics as copper alloys. Nickel 200 (N02200) and alloy B-2 (10665) are resistant but find no reported application. N06600 would be susceptible to pitting in moist CCl_4. The Cr-Ni-Mo alloys should be resistant to the atmospheric boiling point, i.e., from 77°C (170°F) to 100°C (212°F).

Reactive metals

Titanium is resistant to boiling carbon tetrachloride and to a 50% water mixture at ambient temperature and perhaps higher.[4,5] Formation of free HCl would invalidate this data, and the Ti-Pd alloy R52400 would provide more reliable service.

Zirconium is considered fully resistant, but there is no reported usage. Tantalum is also fully resistant.

Precious metals

Silver, gold, and platinum resist boiling carbon tetrachloride, wet or dry, but there are no reported applications.

Other metals and alloys

Lead is resistant to dry CCl_4 to the atmospheric boiling point. Corrosion rates increase in wet solvent, because of the solubility of lead chlorides.

Nonmetallic materials

Organic. The thermoplastic materials are not resistant, except for polyamides at ambient and fluorinated grades at elevated temperatures. Of the thermosetting resins, chlorinated polyesters are resistant at room temperature and epoxy resins to about 66°C (150°F). Above that temperature only those based on phenolic and furane resins are suitable.[6]

Rubbers and elastomers are severely attacked by carbon tetrachloride, except for the fluorinated variety.

Carbon and graphite are resistant and among the preferred materials for coping with wet carbon tetrachloride and its HCl contamination.

Inorganic. Glass and ceramics are fully resistant. Concrete and portland cements are totally unsuitable.

Pitfalls

Corrosion problems may arise from unexpected hydrolysis or catalytic decomposition and HCl attack. Austenitic stainless steels are very susceptible to pitting, intergranular attack, and stress-corrosion cracking. Ferric chloride contamination from corrosion upstream (e.g., in handling chlorine feeds) can aggravate this potential problem.

Mixtures of methanol and carbon tetrachloride can attack aluminum even under conditions in which neither solvent is itself corrosive.[1]

Storage and Handling

Use of the following materials of construction is common for handling and storage and is thought to constitute good engineering practice:

Drums	Galvanized steel (ICC Spec. 17E STC)
Tanks	S30400, A93003, or A95NNN
Tank trucks	S30400, A93003
Railroad cars	S30400, A93003
Piping	S30403
Valves	CF8M
Pumps	CF8M
Gaskets	Spiral-wound PTFE/stainless steel, flexible graphite, expanded PTFE

References

1. I. Mellan, *Corrosion Resistant Materials Handbook,* 3d ed., Noyes Data Corp., Park Ridge, N.J., 1976.
2. *Chemical Age,* vol. 63, no. 6, 1950.
3. C. P. Dillon, "Aluminum in Potentially Corrosive Organic Solvents," *Materials Performance,* vol. 29, no. 11, 1990).
4. B. D. Craig, *Handbook of Corrosion Data,* ASM International, Metals Park, Ohio, 1989.
5. N. E. Hamner, "Corrosion Data Survey," NACE, Houston, 1974.
6. P. A. Schweitzer, *Corrosion Resistance Tables,* 2d ed., Marcel Dekker, New York, 1986.

Ethyl Alcohol

Introduction

Ethanol or ethyl alcohol (C_2H_5OH) has been used for millennia as a beverage and solvent. In modern technology, it is also used as a chemical intermediate (e.g., esters, acetaldehyde, acetic acid). It is a clear, volatile colorless liquid with a pleasant odor. Both anhydrous alcohol (which requires special drying techniques) and the 95% by volume azeotrope (190 proof) are highly flammable; water solutions are not flammable below 50% alcohol content (100 proof).

Process

Historically, and even in modern times, ethyl alcohol has been produced by natural fermentation processes. In industrial production today, the old ethylene-sulfuric acid esterification-hydrolysis process has been supplanted by direct hydration of ethylene, using a phosphoric acid catalyst.

Ethylene is combined with process water and heated at elevated pressure over a catalyst bed to form ethanol (Fig. 43.1). The product mixture of the exothermic reaction is cooled by interchange with the reactor feed. The vapor product is scrubbed with water, vapors recycled, and the combined liquid product distilled to remove light impurities (e.g., ether, aldehyde) and recover a 95% ethanol-water azeotrope. Anhydrous alcohol can be produced from the 95% "spirits" grade by special dehydration techniques (e.g., extraction, azeotropic distillation).

Fermentation of agricultural products (e.g., sugar cane, sugar beets, potatoes, fruit, cellulose, and whey) is an important route in many countries. Water solutions of about 12% alcohol content are obtained, and the 95% alcohol azeotrope produced by distillation.

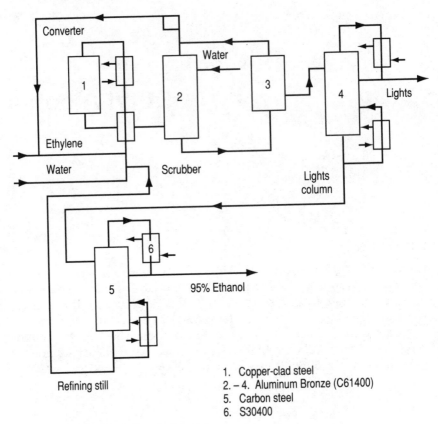

Converter

Water

1

2

3

4   Lights

Ethylene

Water   Scrubber

Lights
column

6

5   95% Ethanol

Refining still

1. Copper-clad steel
2. – 4. Aluminum Bronze (C61400)
5. Carbon steel
6. S30400

Figure 43.1   Ethanol (ethylene hydration).

Materials

Aluminum alloys

Aluminum is generally resistant to ethyl alcohol. Occasional mild pitting is associated with impurities, which vary with the origin of the alcohol. Aluminum is widely used for process equipment, tanks, tank cars, tank trucks, drums, and piping. However, spirits, cosmetic, and low-odor grades are not permitted in aluminum tank trucks. *Anhydrous* alcohol can be severely corrosive at moderately elevated temperatures, evolving hydrogen during the exothermic corrosion, with attendant fire hazard.

Steel and cast iron

Ethanol is not corrosive to iron or steel, but there may be product contamination under small volume-to-area conditions.

Stainless steels

Any type of stainless steel is compatible with ethanol, without reservation.

Copper and its alloys

Copper is the traditional material for distilling and condensing of fine quality ethanol.

Nickel alloys

All nickel-base and nickel-rich alloys are resistant but find no application in this service.

Reactive metals

Titanium, zirconium, and tantalum are unaffected by conventional alcohol. *Anhydrous* ethanol or alcohol with less than 2% water could cause SCC of titanium.[1]

Precious metals

Silver, copper, and gold are fully resistant but find no application in this service.

Other metals and alloys

Lead is fully resistant but finds no application. The resistance of zinc (as galvanizing) is good in straight ethanol, but unsatisfactory in some denatured grades, depending on the specific additive.

Nonmetallic materials

Organic. The usual polyolefin and polyvinyl thermoplastics resist ethanol itself. PE may be unsuitable at moderately elevated temperatures, and all except the fluorinated types may be adversely affected by specific denaturants.

The thermosetting resins are resistant, and baked-phenolic coatings are used in some shipment and storage applications.

All of the synthetic rubbers and elastomers, except polyurethanes, are considered resistant.

Inorganic. The inorganic materials are unaffected by ethanol.

Pitfalls

The only pitfall associated with ethyl alcohol is the attack on aluminum under anhydrous conditions, which fortunately can arise only under deliberate dehydration treatment.

Storage and Handling

Use of the following materials of construction is common for handling and storage and is thought to constitute good engineering practice:

Tanks	Steel, aluminum
Tank trucks	Aluminum (S30400 for spirits, cosmetic, and low-odor grades)
Railroad cars	Aluminum (S30400 as above)
Piping	Aluminum, S30403
Valves	A035600, CF8M
Pumps	CF8M
Gaskets	Neoprene

References

1. I. Mellan, *Corrosion Resistant Materials Handbook,* 3d ed., Noyes Data Corp., Park Ridge, N.J., 1976.

44

Ethyl Benzene

Introduction

Ethyl benzene is of importance primarily as a precursor of styrene (vinyl benzene). Higher homologs are used in manufacture of detergents, and propyl benzene is a cumene precursor. In solvent mixtures, the concentration of ethyl benzene must be limited to 20% because of air pollution considerations.

Process

The primary process is alkylation of benzene with ethylene by one of several processes. One uses aluminum chloride in a homogenous system, while another uses boron trifluoride, and still another (not a Friedel-Crafts reaction) utilizes a molecular sieve (perhaps a crystalline aluminosilicate).

In the widely used Union Carbide/Badger process, alkylation is carried out at low pressure and temperature in a brick- or glass-lined steel reactor, with an aluminum chloride catalyst (Fig. 44.1). The catalyst is recycled from the settler, and the organic phase is washed with water and caustic in the treatment tank, the waste aluminum hydroxide slurry being recovered as class I land-fill material. The crude ethyl benzene is refined, while polyethyl benzenes are recovered in the final column for recycling to the reactor.

The newer molecular-sieve-type process, such as the Mobil/Badger process (Fig. 44.2) is noncorrosive, eliminating the materials selection problems associated with halide-type catalysts. The reaction is carried out in the vapor phase at about 400°C (750°F) over a fixed-bed catalyst. Twin reactors are used, one being regenerated while the alternate is in use. The reaction heat can be used to generate medium-pressure steam. Both low- and high-pressure steam are generated in

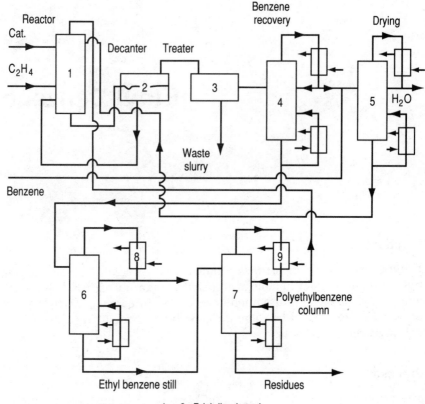

Figure 44.1　Ethylbenzene (Freidel-Crafts reaction).

the still systems during the benzene recovery and ethylbenzene (EB) distillation. Polyethyl benzene (PEB) is recycled to the reactors, and the heavy residues burned as fuel.

Materials

Aluminum alloys

Aluminum is resistant to ethylbenzene, which is reportedly handled in aluminum heat exchangers.[1]

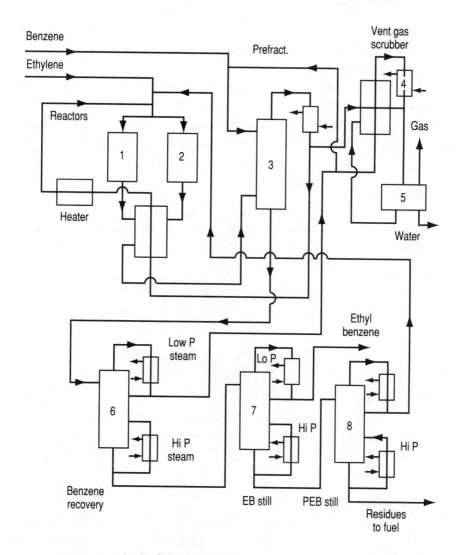

1. – 8. Carbon steel

Figure 44.2 Ethylbenzene (molecular sieve).

Steel and cast iron

Ethylbenzene is intrinsically noncorrosive at room temperature. One source reports 5 to 12 mm/a (20 to 50 mpy) at 100°C (212°F) for steel.[2] However, this probably reflects minor acid contamination in a field exposure.

Stainless steels

Any grade of stainless steel should be resistant to ethyl benzene in the absence of contamination.

Copper and its alloys

Copper and its alloys resist ethylbenzene in the absence of acidity with oxidizing contaminants. However, they are not satisfactory for storage, for unexplained reasons.[3]

Nickel alloys

Nickel-base and nickel-rich alloys are not usually employed in the process and are not required in the absence of specific contaminants.

Reactive metals

Titanium, zirconium, and tantalum are fully resistant but have no application.

Precious metals

Silver, gold, and platinum have no application in this process.

Other metals and alloys

Lead is considered practically resistant with rates of less than 5 mm/a (20 mpy) but has no reported application. Zinc (as galvanized steel) is resistant and considered satisfactory as regards product contamination.

Nonmetallic materials

Organic. Polyethylene is not satisfactory to contain ethyl benzene. By analogy with benzene, polypropylene should be satisfactory to about 60°C (140°F), while thermoplastic materials other than the fluoroplastics above PVDF would not.[4]

FRP construction is not suitable, but reinforced phenolic resins are suitable.[2] An epoxy-phenolic coating is used for tanks and tank cars, but not for over-the-road tank trucks.[1]

None of the rubbers or elastomers, except Viton A, are acceptable.

Pitfalls

All of the above statements regarding behavior of metals and alloys are subject to serious reservations when aluminum chloride catalyst is used. Trace amounts of hydrochloric acid or volatile ferric chloride salts would seriously affect steel, stainless steels, and copper-base alloys particularly.

Storage and Handling

Use of the following materials of construction is common for handling and storage and is thought to constitute good engineering practice:

Tanks	Steel (epoxy-phenolic [EP/PH] if coated), Aluminum
Tank trucks	Aluminum
Railroad cars	Steel [epoxy-phenolic (EP/PH) if coated], Aluminum
Piping	Steel or aluminum
Valves	Steel (6a trim)
Pumps	Steel or cast iron
Gaskets	Graphite braid, Viton A, spiral-wound PTFE/stainless steel

References

1. I. Mellan, *Corrosion Resistant Materials Handbook,* 3d ed., Noyes Data Corp., Park Ridge, N.J., 1976.
2. N. E. Hamner, *Corrosion Data Survey,* 5th ed., NACE, Houston, 1974.
3. Private communication.
4. P. A. Schweitzer, *Corrosion Resistance Tables,* 2d ed., Marcel Dekker, New York, 1986.

Chapter

45

Ethylene Dichloride

Introduction

Ethylene dichloride (EDC; 1,2 dichloroethane) is a colorless volatile liquid with a pleasant odor. It is a solvent (e.g., for fats, greases, and waxes). Major uses are as a starting point for the manufacture of vinyl chloride, other chlorinated solvents (e.g., trichlorethylene, perchlorethylene), and ethylene amines.

Process

EDC is produced by catalytic chlorination of ethylene in a liquid-phase reaction, using a ferric oxide catalyst, at about 50°C (120°F) (Fig. 45.1). Ethylene bromide and acid chlorides (e.g., ferric or cupric chlorides) have also been used as catalysts. A second approach is oxychlorination by HCl and oxygen; the reaction can be effected in a fluidized bed.

Materials

Aluminum alloys

Like other chlorinated hydrocarbons, EDC is potentially corrosive to aluminum.[1] Dry EDC at the boiling point consumed an A92024 coupon in a 7-day test, although addition of 7% water inhibited attack to less than 100 mpy, which is still an unacceptable rate. In general, it is inadvisable to use aluminum alloys in this type of service.[2]

Steel and cast iron

Dry EDC is noncorrosive (less than 1 mpy), but the wet boiling solution gave a rate of about 6 mpy, apparently caused by hydrolysis forming traces of HCl. (*Note:* the asymmetrical isomer, 1,1-dichloroethane,

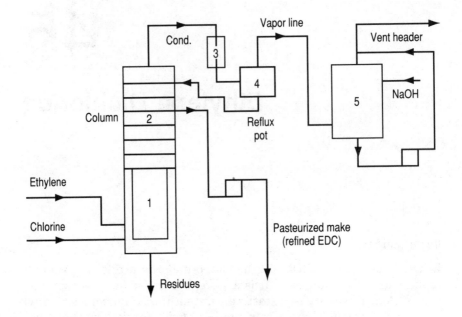

1. Brick-lined base (acid-resistant cement)
2. Steel column (S41000 or S30403 trays);
   N10276 tray support ring
3. Impervious graphite
4. Carbon steel
5. PTFE- or PFA-lined steel

Figure 45.1   Ethylene dichloride.

is more easily hydrolyzed, and causes corrosion problems in distillation systems for purifying EDC.) The refined product can be stored
and shipped in steel.

Stainless steels

Any grade of stainless steel is resistant to dry ethylene dichloride, and
S30400 is approved for storage and shipment.[3] With water at even
moderately elevated temperatures, chloride-induced phenomena such
as pitting, crevice corrosion, and SCC may ensue.[4]

Copper and its alloys

Copper and copper alloys are not approved even for storage of dry EDC
because of chromophoric products. Wet product produces autocatalytic
corrosion due to slow accretion of Cu^{++} ions.

Nickel alloys

Dry EDC is compatible with nickel N02200, while N04400 shows some chromophoric effects due to copper content. Alloy 600 is resistant in dry EDC, but subject to chloride pitting. Alloy N10665 will resist even wet EDC because of its inherent resistance to HCl. The Cr-Ni-Mo alloys (N10276, N06022, N06625) find no application but would be as good as N10665 under wet conditions in the presence of oxidizing contaminants.

Reactive metals

Titanium is shown as resistant to EDC by itself, but will not withstand significant HCl contamination unless oxidizing contaminants are also present. Zirconium and tantalum are resistant.

Precious metals

Corrosion rates for silver are indicated to be less than 20 mpy (5 mm/a).[3] This probably reflects HCl, aeration, and solubilization of silver chloride and should depend on hydrolysis/temperature effects. Gold and platinum are fully resistant.

Other metals and alloys

Galvanized steel and tinned steel are approved for storage and shipment of refined EDC but would be severely affected by traces of HCl contamination.

Nonmetallic materials

Organic. EDC is a powerful solvent, attacking virtually all thermoplastic materials except the fluorinated variety. Of the thermosetting resins, only reinforced phenolic is resistant at 150°C (300°F). Epoxies and polyimides are resistant to about 66°C (150°F).

Only the fluorinated variety of synthetic elastomers is resistant.

Carbon and impervious graphite can be employed to handle EDC contaminated with HCl.

Inorganic. Glass and ceramics are fully resistant.

Pitfalls

Pitfalls are associated with hydrolysis and specific HCl or chloride ion phenomena.

Storage and Handling

EDC is one of the more toxic of the chlorocarbons. Both inhalation and skin contact must be avoided.

Use of the following materials of construction is common for handling and storage and is thought to constitute good engineering practice:

Tanks	Steel (coated with baked phenolic when iron contamination is a concern) or aluminum
Tank trucks	Aluminum or S30400
Railroad cars	S30400
Piping	Steel
Valves	Steel (S41000 trim)
Pumps	Cast iron, steel, or CF8M
Gaskets	Graphite, spiral-wound PTFE/stainless steel

References

1. I. Mellan, *Corrosion Resistant Materials Handbook,* 3d ed., Noyes Data Corp., Park Ridge, N.J., 1976.
2. C. P. Dillon, "Aluminum in Potentially Corrosive Organic Solvents," *Materials Performance,* vol. 29, no. 11, November 1990, p. 51.
3. Private communication.
4. N. E. Hamner, *Corrosion Data Survey,* 5th ed., NACE, Houston, 1974.

Ethylene Glycols

Introduction

Ethylene glycol is a dihydric alcohol [$C_2H_4(OH)_2$], a colorless, odorless high-boiling liquid. Totally miscible with water, it is widely used as antifreeze and as a precursor for linear polyesters for fibers and film.

Process

Ethylene glycol (EG) is produced by hydrolysis of ethylene oxide under substantially neutral conditions (Fig. 46.1). The water effluent is removed in multiple-effect evaporators, the final product being refined by vacuum distillation.

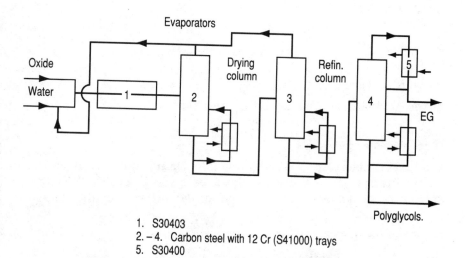

1. S30403
2.–4. Carbon steel with 12 Cr (S41000) trays
5. S30400

Figure 46.1 Ethylene glycol.

Materials

Aluminum alloys

Aluminum and its alloys are resistant to ethylene glycol even at 170°C (338°F).[1] A violent reaction is possible above about 200°C (390°F) under stagnant conditions.[2] This is the typical direct chemical attack of anhydrous alcohols.[3]

Steel and cast iron

Steel and cast iron are not corroded by ethylene glycol or its uncontaminated water solutions.

Stainless steels

Any grade of stainless steel is fully resistant to ethylene glycol.

Copper and its alloys

Copper and its alloys are resistant to ethylene glycol and water solutions. They are not recommended for storage because of chromophoric reactions.

Nickel alloys

Nickel and its alloys are fully resistant.

Reactive metals

Titanium, zirconium, and tantalum are fully resistant, but find no application in this service.

Precious metals

Silver, gold, and platinum are fully resistant.

Other metals and alloys

Lead is described as corroding at less than 5 mm/a (20 mpy) in 50% at ambient temperature and 100% glycol up to 100°C (212°F).[1,4] Neither galvanized steel nor tinned steel is permitted in shipment and storage.

Nonmetallic materials

Organic. All conventional plastics and elastomers are resistant to ethylene glycol within normal temperature limitations.

Inorganic. Glass and ceramics are completely resistant to ethylene glycol and its water solutions.

Pitfalls

Aside from reaction with aluminum at elevated temperatures, the only pitfalls are those associated with contamination. Glycol slowly oxidizes at elevated temperatures to form trace amounts of glycolic acid. For this reason, ethylene glycol antifreeze is buffered and inhibited.

Storage and Handling

Use of the following materials of construction is common for handling and storage and is thought to constitute good engineering practice:

Tanks	Steel, aluminum, or S30400
Tank trucks	Steel, aluminum, or S30400
Railroad cars	Steel, aluminum, or S30400
Piping	Steel, aluminum, or S30400
Valves	A03560, CF8M
Pumps	A035600, steel, or cast iron
Gaskets	Any

References

1. I. Mellan, *Corrosion Resistant Materials Handbook*, 3d ed., Noyes Data Corp., Park Ridge, N.J., 1976.
2. B. D. Craig, *Handbook of Corrosion Data*, ASM International, Metals Park, Ohio, 1989.
3. C. P. Dillon, "Aluminum in Potentially Corrosive Organic Solvents," *Materials Performance*, vol. 29, no. 11, November 1990.
4. N. E. Hamner, *Corrosion Data Survey*, 5th ed., NACE, Houston, 1974.

47

Ethylene Oxide

Introduction

Ethylene oxide is a colorless gas. which can be condensed to liquid form at low temperatures, b.p. 10.7°C (51°F). It is miscible with water and many organic solvents. The precursor of ethylene glycol, it is also used in the manufacture of glycol ethers, ethanolamines, and surfactants.

Process

Ethylene is oxidized at approximately 200° to 300°C (390° to 570°F) over a silver catalyst at 1 to 3 MPa (145 to 435 lb/in^2) in steel tube-and-shell converters (Fig. 47.1). The crude oxide is scrubbed with cold water, unreacted ethylene, and carbon dioxide passing overhead. The oxide is desorbed under reduced pressure and distilled and compressed for storage in liquid form under nitrogen.

Materials

Aluminum alloys

Aluminum and its alloys are fully resistant to ethylene oxide to, conservatively, 100°C (212°F).[1] Its water solutions may cause pitting if certain impurities enter with the water.[2]

Steel and cast iron

Steel and cast iron are unattacked by ethylene oxide, but the latter material is not used because of the danger from brittle fracture. For cold-temperature storage, special fine-grained killed steels with good

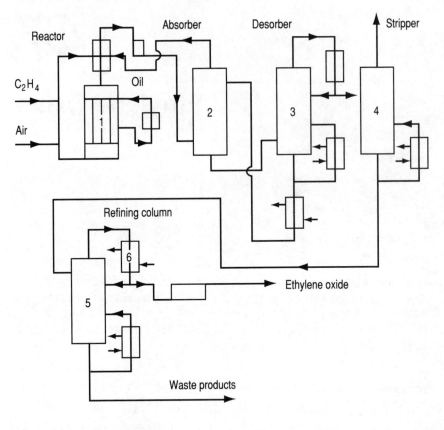

1. S30403
2. – 5. Carbon steel columns with 12 Cr (S41000) trays
6. S30400

Figure 47.1   Ethylene oxide.

nil-ductility transition temperature characteristics are specified (e.g., ASTM A516-70).

Stainless steels

All grades of stainless steels are resistant. S30400 or S30403 (which is not really required) is the conventional selection for ease of fabrication and excellent low-temperature properties.

Copper and its alloys

Copper and copper alloys are unaffected by ethylene oxide. However, the U.S. Department of Transportation (DOT) precludes them from use in tank cars—a hangover from concerns about acetylide formation in oxide produced by previous processes.

Nickel alloys

Nickel alloys are fully resistant but find no application in this service. Electroless nickel-plated vessels can be used to maintain purity.

Reactive metals

There is no occasion to use titanium, zirconium, or tantalum in this service.

Precious metals

Silver, gold, and platinum find no application in this service. Silver would be precluded in the presence of acetylene contamination.

Other metals and alloys

Lead is reported to be practically resistant, being attacked at less than 20 mpy (0.5 mm/a).[1,2] Zinc as galvanizing is resistant but is unsatisfactory because it catalyzes polymerization.[3] There appears to be no corresponding data on tinned steel.

Nonmetallic materials

Organic. Of the organic plastics, only fluorinated plastics or elastomers, and polyamides, are unaffected by the strong solvent action of ethylene oxide. Even so, FEP and PTFE are rated to not more than about 93°C (200°F) and the CTFE and ECTFE to no more than 38°C (100°F).[4]

A polyester copolymer material is reportedly resistant and a polyester elastomer, Hytrel[TM], is considered resistant.[2]

Carbon and graphite are resistant, but the resin-cemented joints require phenolic, furane, sulfur, or acid-resistant silicate mortars.

Inorganic. Glass and ceramicware is resistant but is not employed because of the possible hazard of mechanical damage.

Pitfalls

Ethylene oxide may undergo slow polymerization in storage. This is aggravated by high temperatures and by contamination with water, alkalies, acids, metal oxides, etc. with possible pressure excursion or explosion.

Storage and Handling

Refined ethylene oxide is stored in gaseous form under about 618 kPa (75 lb/in^2) gauge pressure. The gas will cause eye irritation, and the material is relatively toxic as well as flammable and explosive, particularly in air or oxygen mixtures.

Use of the following materials of construction is common for handling and storage and is thought to constitute good engineering practice:

Tanks*	Steel (A516-70)
Tank trucks	Steel
Railroad cars	Steel
Piping	Steel
Valves	Steel (S41000 trim) or CF8
Pumps	Steel or CF8
Gaskets	Spiral-wound PTFE/stainless steel, graphite fiber

References

1. N. E. Hamner, *Corrosion Data Survey*, 5th ed., NACE, Houston, 1974.
2. I. Mellan, *Corrosion Resistant Materials Handbook*, 3d ed., Noyes Data Corp., Park Ridge, N.J., 1976.
3. Private communication.
4. P. A. Schweitzer, *Corrosion Resistance Tables*, Marcel Dekker, New York, 1986.
5. B. A. Martin, "Corrosion Protection of Mounded LPG Tanks," *Materials Performance*, vol. 29, no. 9, September 1990, p. 13.

Note: Tanks must be well-insulated, protected by water-spray systems, equipped with cooling coils, and electrically grounded. Alternatively, they may be mounded and cathodically protected.[5]

Chapter

48

Fatty Acids and Tall Oil

Introduction

Tall oil (a black, sticky, viscous liquid composed mostly of resin and fatty acids) is a natural product of pine trees . Fatty acids are derived by acidification of the soapy product from wood pulping operations. Tall oil contains ether-soluble nonlignin, noncellulosic compounds used as paper-making additives and in rubber and adhesives, while the tall oil fatty acids (TOFA) are mainly 18-carbon acids (e.g., oleic, linoleic acids) produced by distillation. TOFAs find application in intermediate chemicals, protective coatings, soaps and detergents, and similar products.

Process

Tall oil is a by-product of Kraft pulping of pine trees, alkaline digestion delignifying the wood to form cellulose pulp, sodium soaps, salts of fatty acids, and lignin degradation products. Black liquor is concentrated, skimmed, and acidified, producing crude tall oil, an interfacial layer (oil, pulp fines, calcium sulfate) and a waste-acid layer (Fig. 48.1).

Black-liquor soap is collected (it may be solvent-extracted or brine-washed) and acidified with 30 to 40% sulfuric acid, producing sodium sulfate and fatty acids (RCOOH). Reactor products are decanted (waste brine recycled or discarded) and crude products washed with hot water or brine. Distilled tall oil is made overhead from a short tower, without fractionation, while crude tall oil is distilled overhead in the rosin tower. The TOFA crude product is then preheated and refined, noncondensables being vented, while high-boiling contaminants are discarded as pitch residues from the base.

TOFAs may be further fractionated and refined (including various

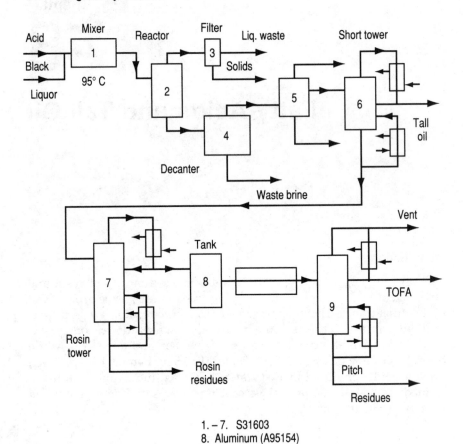

Figure 48.1   Tall oil/fatty acids.

1.–7. S31603
8. Aluminum (A95154)
9. S31603

chemical, physical, and clay-bleaching operations) to produce proprietary products.

Materials

Aluminum alloys

Aluminum and its alloys resist fatty acids and are used for process
equipment and storage tanks.[1] *Anhydrous* acids at about 250°C
(480°F) become severely corrosive.[2,3] However, a small amount of water (e.g., 1000 ppm) is sufficient to inhibit the corrosion.

Steel and cast iron

Steel is indicated to be unsatisfactory.[4] More than likely, this is a result of iron contamination.

Unalloyed cast iron can be used to handle concentrated fatty acids, but suffers attack in dilute aqueous solutions.[3] The nickel cast iron F43000 is about an order of magnitude more resistant, while the silicon iron F47003 is inert.[1]

Stainless steels

Any grade of stainless steel is resistant at ambient temperatures. The molybdenum-free 18-8 grades are definitely unsatisfactory at even moderately elevated temperatures, but S31603 is usually resistant to about 200°C (390°F).[3] However, rates of 5 to 8 mm/a (20 to 30 mpy) may be encountered with small amounts of sulfuric acid contamination and N08020 or similar alloys are then required.[5]

Copper and its alloys

Copper can be quite resistant even at elevated temperatures, although discoloration may be a problem.[2] The cupronickels may be more resistant. However, all copper alloys are attacked in the presence of moisture and air.[3]

Nickel alloys

Alloy 400 (N04400) has been used in some saponification processes. Otherwise, the lower nickel-base alloys usually find little or no application. The Ni-Mo (N10665) and Ni-Cr-Mo alloys (N10276, N06625, N06022) are resistant.

Reactive metals

Titanium, zirconium, and tantalum are shown as resistant.[2] However, traces of sulfuric acid could be potentially dangerous to titanium.

Precious metals

Silver, gold, and platinum are fully resistant to 400°C (750°F) in the fatty acids themselves.[3]

Other metals and alloys

Lead and zinc are generally unsuitable, the former being adversely affected by dissolved oxygen.[1,2,4] Tin is resistant at room temperature but becomes nonresistant at moderately elevated temperatures, e.g., over 60°C (140°F).[3]

Nonmetallic materials

Organic. Conventional thermoplastic and thermosetting resins are resistant at room and moderately elevated temperatures.[4] The FRPs' limiting service temperatures vary from about 66°C (150°F) to 93°C (200°F), depending on the specific formulation. The fluorinated varieties may be used at elevated temperatures to their conventional limits.

Of the rubbers and elastomers, only natural rubber and neoprene are deemed unsuitable at room temperature.

Furane and silicate mortars are resistant, while sulfur cements are not recommended.[1]

Unlined wooden tanks are reportedly unattacked in fatty acids, and carbon and graphite are unaffected.

Inorganic. Glass and ceramicware is unaffected by TOFAs.

Pitfalls

The only pitfalls related to TOFA service are those associated with contamination (e.g., sulfuric acid, halides) which can have specific adverse effects.

Storage and Handling

Use of the following materials of construction is common for handling and storage and is thought to constitute good engineering practice:

Tanks	Aluminum or S30400
Tank trucks	Aluminum or S30400
Railroad cars	Aluminum or S30400
Piping	Aluminum or S30400
Valves	A035600 or CF8M
Pumps	CF8M or CN7M
Gaskets	Buna N, spiral-wound PTFE/stainless steel

References

1. I. Mellan, *Corrosion Resistant Materials Handbook*, 3d ed., Noyes Data Corp., Park Ridge, N.J., 1976.
2. N. E. Hamner, *Corrosion Data Survey*, 5th ed., NACE, Houston, 1974.
3. B. D. Craig, *Handbook of Corrosion Data*, ASM International, Metals Park, Ohio, 1989.
4. P. A. Schweitzer, *Corrosion Resistance Tables*, Marcel Dekker, New York, 1986.
5. J. D. Polar, *A Guide to Corrosion Resistance*, Climax Molybdenum Co., New York, 1981.

49

Fluorocarbons

Introduction

Fluorocarbons are compounds of carbon, fluorine, and chlorine containing little or no hydrogen. They are chemically inert and very stable and have been used as refrigerants, blowing agents, solvents, and precursors to fluoropolymers.

Despite their inertness, they do react with ozone. The fluorocarbon gases have been implicated in depletion of the upper-atmosphere ozone layer which shields the earth from ultraviolet rays. Many are being phased out of production. However, those which contain hydrogen (e.g., propellants 22, 142b) are claimed to be destroyed in the lower atmosphere, while those which do not contain chlorine (e.g., propellant 152a) are reportedly not implicated in the ozone layer depletion.

Process

The fluorocarbon gases have been made by reaction of carbon tetrachloride (propellants 11, 12) or chloroform (propellant 22) with anhydrous hydrogen fluoride in the presence of a catalyst such as molten antimony pentachloride.[1] The liquid-phase process is shown in Fig. 49.1. Corrosion is more of a problem with propellant 22, because it is more easily hydrolyzed than are propellants 11 and 12.

The materials of construction are only suggestions; process changes and varying conditions may profoundly affect materials performance (see "Pitfalls" below).

Details of processes for the newer, more environment-compatible propellants are not currently available.

Materials

Aluminum alloys

Aluminum and its alloys are resistant to dry fluorocarbon gases.[2] However, the gases can react with moisture, forming traces of hydro-

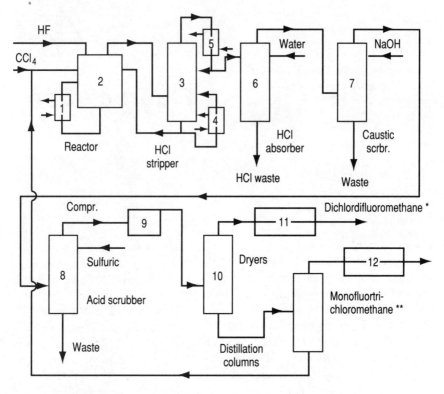

HF

CCl₄

2   Reactor

1

3   HCl stripper

5

4

6   HCl absorber

Water

HCl waste

7   Caustic scrbr.

NaOH

Waste

Compr.

9

Sulfuric

8   Acid scrubber

Waste

10   Dryers

11

Distillation columns

Dichlordifluoromethane *

12

Monofluortri-chloromethane **

* Propellant 12   ** Propellant 11

1. 80 Cu-20 Ni (C71000)
2. S30403 (C71000 in velocity areas)
3. Aluminum-killed steel (K11267)
4. Ni-Cr-Mo (N10276)
5. K11267 steel (refrigerated)

6. Brick-lined steel
7. Alloy 400 (N04400)
8. Cr-Ni-Fe-Mo-Cu (N08825 or N08020)
9. – 12. Carbon steel

Figure 49.1 Fluorocarbons.

chloric and hydrofluoric acid which attack the metal. The use of aluminum alloys in gas refrigerant systems is contingent on absolutely anhydrous conditions.[3]

Steel and cast iron

Steel and cast iron are considered unsuitable for the fluorocarbon gases by one source.[4] This is probably on the basis of NDTT and inherent brittleness respectively, since another authority indicates steel as excellent.[2]

Stainless steels

Any grade of stainless steel is resistant to dry fluorocarbons. The austenitic 18-8 stainless grades are preferred for ease of fabrication and good low-temperature properties, while the molybdenum-bearing alloy S31603 is preferred because of a slight advantage in the event of halide contamination.

Copper and its alloys

Copper and its alloys are resistant to dry fluorocarbon gases.[5] The 80-20 cupronickel (C71000) has proven superior in the liquid-phase reaction.

Nickel alloys

Alloy 400 (N04400) is rated excellent in refrigerant 11 and good in refrigerant 12 to about 200°C (390°F), but only at room temperature in refrigerant 22.[4] Nickel alloy 200 is not as good, while there is little or no occasion to consider the higher nickel-base alloys except in some process streams.

Reactive metals

There is no occasion to use titanium, zirconium, or tantalum in this service. Indeed, traces of fluoride contamination would be highly inimical.

Precious metals

There is no application for silver, gold, or platinum in handling the refined product. Both fine silver (P07015) and sterling silver (P07931) have been used with some success in process equipment.

Other metals and alloys

Lead is moderately resistant at ambient temperature but finds no practical application.

Nonmetallic materials

Organic. Over the broad range of fluorocarbons, only the fluorinated plastics have a wide range of resistance. There is little occasion to use either thermoplastic or thermosetting resins in fluorocarbon products.

Resistance of elastomers varies from excellent to unusable, depending on the specific fluorocarbon. Neoprene is the one most nearly resistant to all the common refrigerants.[2]

Impervious graphite heat exchangers are approved for fluorocarbon gases at ambient temperature.

Inorganic. Glass and ceramicware are resistant to anhydrous fluorocarbons. Traces of hydrofluoric acid would be corrosive.

Pitfalls

The two pitfalls in handling fluorocarbon products are (1) the adverse effects of free HCl or HF liberated by hydrolysis upon water contamination and (2) the influence of oxidizing cations as contaminants (e.g., Sb^{+5}) in the process. All heat exchangers carrying water or steam vs. fluorocarbons should have welded tube-to-tubesheet joints.

In one process, pilot plant operation was a great success in austenitic stainless steels but a disaster in full-scale operation, with severe pitting, SCC, and even general corrosion occasioned by the degradation products and cation contaminants.

Storage and Handling

Use of the following materials of construction is common for handling and storage and is thought to constitute good engineering practice for handling refined fluorocarbon products under anhydrous conditions:

Tanks	Aluminum or S30400
Tank trucks	Aluminum or S30400
Railroad cars	Aluminum or S30400
Piping	Aluminum or S30400
Valves	A03560 or CF8M
Pumps	CF8M
Gaskets	Neoprene, graphite, spiral-wound PTFE/stainless steel

References

1. G. T. Austin, *Shreve's Chemical Process Industries,* 5th ed., McGraw-Hill, New York, 1984.
2. I. Mellan, *Corrosion Resistant Materials Handbook,* 3d ed., Noyes Data Corp., Park Ridge, N.J., 1976.
3. C. P. Dillon, "Corrosion of Aluminum by Organic Solvents," *Materials Performance,* vol. 29, no. 11, November 1990.
4. P. A. Schweitzer, *Corrosion Resistance Tables,* Marcel Dekker, New York, 1986.

Formaldehyde

Introduction

Formaldehyde is a colorless gas with a clear pungent odor. It is sold as an aqueous solution of 30 to 50% concentration; the 37% and 44% grades (formalin) may be inhibited with 5 to 8% methanol, while 50% and 56% products are uninhibited low-methanol (less than 2%) solutions. Formaldehyde is a basic building block in chemicals ranging from resin products to fertilizers.

Process

The predominant process is oxidation (and simultaneous dehydrogenation) of methanol at about 625°C (1150°F) at 34 to 69 kPa (5 to 10 lb/in^2) pressure over a silver catalyst (Fig. 50.1). The product is rapidly cooled in a steam generator and a water-cooled condenser, and enters an absorption tower. The absorber bottoms feeds a distillation tower, where methanol is recovered and recycled. A 55% formaldehyde solution is recovered as a tails product. Type 316L stainless steel is required both to resist corrosion by traces of formic acid and to protect the catalyst from iron contamination.

A metal oxide catalyst has also been used, reducing the reaction temperature to 300 to 400°C (575 to 750°F).

Materials

Aluminum alloys

Aluminum and its alloys resist the dry formaldehyde gas and the methanol-inhibited water solutions. However, traces of formic acid or benzaldehyde as impurities may cause pitting.[1]

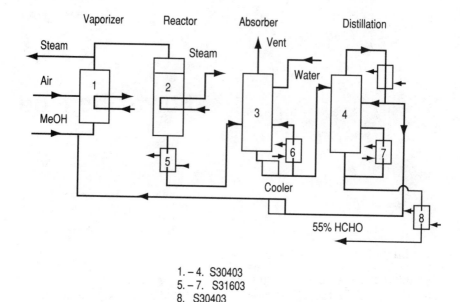

1. – 4. S30403
5. – 7. S31603
8. S30403

Figure 50.1   Formaldehyde.

Steel and cast iron

Although resistant to the dry gas, steel and cast iron are deemed un-suitable because even small amounts of moisture cause corrosion. They are definitely unsuitable for the formalin solutions. Nickel cast iron (F43000) is an order of magnitude more resistant, but there seems little occasion to use it. Silicon cast iron (F47003) is resistant but is not employed.

Stainless steels

All grades of stainless steel resist the pure formaldehyde. In formalin, S30400 is suitable for ambient temperature. At elevated tempera-tures, S31603 is preferred to resist traces of formic acid. The high-performance stainless steels, such as N08020 and N08007, are even more resistant in the presence of acid contamination.

Copper and its alloys

With the exception of the high-zinc yellow brasses, copper and its al-loys are considered fully resistant to formaldehyde.[1] If formic acid is present, corrosion is proportional to oxygen or oxidizing agents, including the accretion of cupric ion corrosion products. Copper is approved for storage and shipment.

Nickel alloys

Nickel N02200 and N04400 are excellent in dilute solutions. The nickel-base alloys find little application in formaldehyde and formalin applications, and nickel is not approved for storage and shipment.

Reactive metals

Titanium will resist boiling 37% formaldehyde solutions.[1] Zirconium and tantalum are likewise resistant.

Precious metals

Silver, gold, and platinum resist all concentrations to the boiling point but find no application in this service.

Other metals and alloys

Lead resists 100% formaldehyde and corrodes at less than 0.5 mm/a (20 mpy) in formaldehyde solutions, but is not recommended for this service.[1] Tin is resistant to about 60°C (140°F), but a steel substrate would be susceptible to attack at pinholes in tinplate. Zinc and tin are not permitted for storage or shipment.

Nonmetallic materials

Organic. Many thermoplastic materials resist formaldehyde solutions, especially at ambient temperature. Polyethylene should not be exposed above about 60°C (140°F), but polypropylene is resistant to about 80°C (175°F). PVC is satisfactory at 60°C (140°F) also. PVDF is more limited than one would expect, perhaps to 120°C (250°F). FEP and the other fluoroplastics are suitable at appropriate elevated temperatures.

Reinforced epoxy, phenolic, and polyester resins will withstand formalin, at least to about 75°C (170°F).

Butyl rubber is recommended for formalin, and butadiene styrene and nitrile rubbers are resistant, but neoprene is not above ambient temperature.

Furane, polyester, silicate, and sulfur mortars are satisfactory, and phenolic cement may be used to 177°C (350°F).

Carbon and graphite are unaffected by formalin.

Inorganic. Glass and ceramicware are totally resistant to formaldehyde solutions.

Pitfalls

The pitfalls associated with formaldehyde and its aqueous solutions are the possibility of formic acid contamination due to oxidation, the possible influence of halides on stainless steels, and the adverse effects of halides and/or heavy metal contamination on aluminum.

Storage and Handling

Use of the following materials of construction is common for handling and storage and is thought to constitute good engineering practice:

Tanks	Aluminum or S30400
Tank trucks	Aluminum or S30400
Railroad cars	Aluminum or S30400
Piping	Aluminum or S30400
Valves	A03560, CF8M
Pumps	CF8M
Gaskets	Butyl rubber, graphite fiber, spiral-wound PTFE/stainless steel

References

1. I. Mellan, *Corrosion Resistant Materials Handbook,* 3d ed., Noyes Data Corp., Park Ridge, N.J., 1976.

51

Formic Acid

Introduction

Formic acid is the simplest aliphatic acid, H · COOH, with a melting point of 8.4°C (47°F). A colorless, odorous liquid, it is a strong acid and, in the anhydrous state, a powerful dehydrating agent. The strong acid has a tendency to decompose, releasing carbon monoxide and water. Its azeotrope with water is anomalous, the 77.5% acid having a boiling point *higher* than its constituents (b.p. 107.3°C). Formic acid is used as a preservative for cattle fodder and in various chemical processes (ester manufacture, rubber coagulation, etc.).

Process

Modern syntheses have largely replaced the older process, which involved reaction of caustic with carbon monoxide followed by hydrolysis of sodium formate with dilute sulfuric acid. In Europe, it is produced by oxidation of formamide.

Formic acid is produced as a by-product of acetic acid manufacture by butane oxidation (Chap. 33). About one-third of the world's production is from methyl formate and ammonia via formamide, which hydrolyzes with 72% sulfuric acid to produce formic acid and ammonium sulfate.

The Leonard process (Fig. 51.1) produces methyl formate by catalyzed reaction of steam and CO, followed by hydrolysis to methanol and formic acid under high pressure. The methanol is recovered, while the formic acid is concentrated in a zirconium distillation system.

Materials

Aluminum alloys

Aluminum and its alloys are attacked by formic acid above room temperature and below about 20 to 30%.[1] Although a protective film is

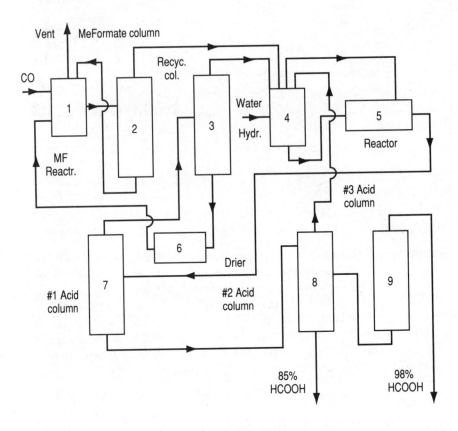

Figure 51.1   The Leonard process.

formed at higher concentrations, aluminum is not recommended for this service except in special applications. Contaminants, especially chlorides or heavy metal salts, are a particular problem. Aluminum has sometimes been used for shipment of 95 to 99% concentration.[2]

Steel and cast iron

Steels are unsuitable for formic acid exposure, as are the cast irons, excepting only the 14% silicon variety (F47003). The latter is excellent in water solutions, but its rating drops off in anhydrous acid above ambient temperature.[3]

Stainless steels

S30403 is the preferred material for storage and shipment of formic acid (80 to 100%). However, S31603 is required at even slightly elevated temperatures. The duplex alloys are superior even to S31703.[4] The superferritics (e.g., S44627) are useful in less than 80% formic acid at the boiling point, and nickel-rich grades (e.g., N08020) are preferred at elevated temperatures and concentrations.

Copper and its alloys

With the exception of the high-zinc brasses, all copper alloys are resistant to formic acid under strictly anaerobic and nonoxidizing conditions. Copper-clad steel and aluminum bronze are widely used in formic acid processes. However, in the presence of oxidants of any kind, they are rapidly corroded.

Nickel alloys

The nonmolybdenum grades (N02200, N04400, N10665) are resistant under nonoxidizing conditions, but are not economically competitive with copper alloys.

The Ni-Cr-Mo alloys (N06625, N10276, N06022) may be used at elevated temperatures and in the presence of contaminants which render stainless-type alloys unsuitable.

Reactive metals

Titanium can be highly resistant, particularly if oxidizing contaminants are present (e.g., Fe^{+++}, Cu^{++}). However, it is attacked at catastrophic rates in boiling anhydrous acid.

Zirconium is excellent up to 90% acid, but resistance diminishes with less than 10% water above about 50°C (125°F).[1] Tantalum is fully resistant to all concentrations to at least the atmospheric boiling point.

Precious metals

Silver is attacked above about 50% concentration, but platinum is fully resistant.

Other metals and alloys

Lead, tin, and zinc are not resistant in this service.

Nonmetallic materials

Organic. Formic acid is both an acid and an aldehyde. The solvent characteristic renders many plastics and elastomers unsuitable. Polyethylene is not approved for shipping containers, although nominally resistant. Polypropylene is useful to about 50°C (120°F). Only the fluorinated plastics should be used at elevated temperatures.

Specific thermosetting resins are resistant at specific concentrations to definite temperature limitations. The manufacturer should be consulted. Formic acid is a specific solvent for many nylon formulations.

Of the rubbers and elastomers, neoprene is useful at ambient temperature but fluorinated elastomers are more resistant, with a temperature limit of about 50°C (120°F). Chlorsulfonated polyethylene is suitable.

Furane, furane/epoxy, phenolic, and silicate cements, but not sulfur cement, are resistant.

Carbon and impervious graphite resist all concentrations of formic acid.

Inorganic. Glass and ceramics are fully resistant, but portland cement and concrete are rapidly attacked.

Pitfalls

In prolonged storage above about 30°C (85°F), formic acid slowly decomposes with liberation of CO. Although 99.5% pharmaceutical grade is less than 5 ppm Cl^-, 90 and 98% technical grades can contain up to 20 ppm.

Oxidizing contaminants have a catastrophic effect on copper alloys, while halide contamination causes localized (or even general) corrosion of austenitic stainless steels.

A titanium-tubed condenser, selected to resist seawater and intermediate-strength formic acid operated successfully for some period of time, but was destroyed quickly during a start-up operation utilizing 100% acid.

Storage and Handling

A DOT white label is mandatory. Because 98 to 99% acid (e.g., technical and pharmaceutical grades) has a higher coefficient of thermal expansion than 90% acid, provisions must be made to accommodate this.

Use of the following materials of construction is common for handling and storage and is thought to constitute good engineering practice:

Tanks   S30403 (N08020 heating coils)

Tank trucks   S30403

Railroad cars	S30403
Piping	S31603
Valves	CF3M. CN7M
Pumps	CN7M
Gaskets	Graphite fiber, spiral-wound PTFE/stainless steel

References

1. I. Mellan, *Corrosion Resistant Materials Handbook*, 3d ed., Noyes Data Corp., Park Ridge, N.J., 1976.
2. B. D. Craig, *Handbook of Corrosion Data*, ASM International, Metals Park, Ohio, 1989.
3. P. A. Schweitzer, *Corrosion Resistance Tables*, Marcel Dekker, New York, 1986.
4. J. D. Redmond, "Selecting Second-Generation Duplex Stainless Steels," *Chemical Engineering*, October 27, 1986.

52

Hydrogenation

Introduction

In this chapter, we digress somewhat from the usual style and format. There are simply too many types of hydrogenations to be summarized herein. Nevertheless, there are certain peculiarities about hydrogenation processes which need to be brought to the attention of the process engineer.

Process

Hydrogenation is the process of reacting hydrogen with another molecule. The hydrogenation of unsaturated organics produces the corresponding saturated chemical. Hydrogenation reactions are used to manufacture citronellol from citronellal, to hydrogenate oils and greases (partially or totally) for both soap and food industries, and to produce isooctane from diisobutylene.[1] Besides saturating double bonds, hydrogenation can be employed to eliminate such elements as oxygen, nitrogen, halogens, and sulfur from a chemical compound.

Hydrogenations are effected catalytically at elevated temperatures and pressures. Various metals and their oxides are effective catalysts. When hydrogen is produced by the interaction of hydrocarbons and steam, such catalyst poisons as sulfur compounds and carbon monoxide must be effectively removed from the hydrogen gas.

Materials

When hydrogen is produced by corrosion (as by acid corrosion of metals below copper in the electromotive force series or by alkaline corrosion of amphoteric metals), the atomic hydrogen usually dimerizes and is evolved in the harmless molecular form. However, if the dimerization is slowed or prevented by specific contaminants (notoriously sulfides, but also selenides, arsenides, cyanides, and antimony compounds), the atomic hydrogen penetrates the metallic crystal structure. It may then cause a transient loss of ductility or even cause blistering or environmental cracking.[2]

At elevated temperatures, e.g., greater than 230°C (450°F), there is *always* some atomic hydrogen in equilibrium with the molecular hydrogen. This will also cause some degree of embrittlement, depending on the specific metal, as well as internal reactions with intermetallic compounds to produce reaction products within the interstices of the metal structure. Further, under conditions of low strain rate and high-purity, high-pressure hydrogen, there may be a degradation of mechanical properties.[3]

Aluminum alloys

Dry hydrogen gas is not inimical to aluminum alloys, but, with increasing humidity, subcritical crack growth can occur. This is observed primarily in the 7000 series (Al-Zn-Mg) and is probably not a problem for the alloys used in process industries.

Steel and cast iron

Cast iron is not suitable for high-pressure hydrogen. Hydrogen attack on steel commences above about 200°C (400°F) at about 13.8 MPa (2000 lb/in^2) *partial* pressure of hydrogen, with internal decarburization producing methane in the interstices of the metal. Such attack has been reported also at 1.4 MPa (200 lb/in^2) at about 315°C (600°F).[4] Traditionally, the Nelson curves (API publication no. 949, latest edition) define the relationships between temperature and partial pressure above which alloy steels are used.[5] Increasing the chromium content of alloy steels containing about 0.5% molybdenum from about 1 to 9% produces stable carbides resisting chemical reaction at higher temperatures and pressures. Ultimately, 18-8 stainless steels are required.

Stainless steels

Martensitic stainless steels are susceptible to hydrogen embrittlement. The ferritic and austenitic grades are not, *except* when heavily cold-worked. In that condition, martensite is formed by deformation. There is also a possible loss in ductility in high-pressure hydrogen, with the less stable S30403 and S34700 suffering up to 50 to 60% loss of reduction in area, while more stable austenitic varieties (S31000, S31603, S30900) are substantially unaffected.

Copper and its alloys

At elevated temperatures above the critical temperature for water, copper and copper alloys containing either oxygen in solid solution or as oxide inclusions suffer internal fissuring due to localized steam generation. This is primarily a problem in bright annealing copper in a hydrogen atmosphere.

Nickel alloys

Cold work and/or aging can produce reduced ductility in nickel-based alloys. The austenitic types are less susceptible. These alloys are not usually utilized except as corrosion-resistant nickel, N06600, or Ni-Cr-Mo *cladding* for hydrogenation vessels in specific processes, in which case it is the *backup* steel substrate which is of concern (see "Pitfalls" below).

Reactive metals

Titanium, zirconium, and tantalum are all reactive with hydrogen and probably would not be employed even in clad-steel construction. Titanium and zirconium actually exhibit ductile-to-brittle transition characteristics related to hydrogen content. Adsorption of hydrogen by titanium is reduced by water content in the range of 2%. Tantalum is generally nonreactive below about 250°C (480°F) but will absorb at room temperature under deformation if it contains oxygen.

Precious metals

Platinum is quite susceptible to hydrogen embrittlement. There is little occasion to use precious metals in hydrogen service.

Other metals and alloys

Lead, zinc, and tin have no application in hydrogen service.

Nonmetallic materials

Nonmetallic materials are not usually considered for hydrogenation service.

Pitfalls

Leakage of high-pressure hydrogen is very dangerous because of the flammability and explosive nature of hydrogen-air mixtures. In addition, unlike conventional gases which are heated in compression and cooled by expansion, hydrogen gets hot on expansion. It can be self-igniting, and the almost invisible flame jet is a hazard to personnel as well as adjacent structures.

When hydrogenations are conducted in molten sodium environments, trace amounts of caustic may form from parts per million of contained moisture. SCC of alloy or stainless steels may ensue, with an attendant high-pressure hydrogen leak. At elevated temperatures, traces of water are formed by reduction of metallic oxides, as indicated above.

Almost all metals and alloys are highly permeable to *atomic* hydrogen. Consequently, alloy steel vessels integrally *clad* with corrosion-resistant alloys like copper, nickel, or stainless steel must be of the proper alloy content themselves (see "Steel and cast iron" above) to be unaffected at the operating temperature and pressure. Rolled or multilayer vessels permit hydrogen to escape between the layers, rather than reacting with the steel itself.

Storage and Handling

Use of the following materials of construction is common for handling and storage and is thought to constitute good engineering practice:

Tanks	Steel
Tank trucks	Steel
Railroad cars	Steel
Piping	Steel
Valves	Steel
Gaskets	Metallic

References

1. G. T. Austin, *Shreve's Chemical Process Industries,* 5th ed., McGraw-Hill, New York, 1984.
2. C. P. Dillon, *Corrosion Control in the Chemical Process Industries,* McGraw-Hill, New York, 1986.
3. "Corrosion," *Metals Handbook,* 9th ed., vol. 13, ASM International, Metals Park, Ohio, 1987.
4. "Steels for Hydrogen Service at Elevated Temperatures and Pressures in Petroleum Refining and Petrochemicals," API publication no. 941, API, Washington, D.C., 1990.
5. G. R. Prescott, "Material Problems in the Hydrocarbon Processing Industries," *Alloys for the Eighties,* Climax Molybdenum Co., New York.

Phenol

Introduction

Phenol (C_6H_5OH), or *carbolic acid,* is a colorless, white crystalline compound with a characteristic sweet aromatic odor. It is used in extraction of lubricating oils, production of resin molding materials, manufacture of antiseptics, and other processes.[1] Technical grades of phenol melt at 42.8°C (109°F) and boil at 182.8°C (361°F), while the freezing point is 41.1°C (106°F). Liquid phenol is made by melting the crystals and adding water (71.28 g per 100 g phenol) to give a product of 58.4% carbolic acid which is liquid at room temperature. Having a benzene ring structure, phenol is soluble in most organic solvents and is soluble in oil to an extent depending on the specific composition of the oil.

Phenol is itself only mildly corrosive to steel, and noncorrosive to aluminum, below about 150°C (300°F) in the presence of traces of water (0.3% minimum).[2] However, discoloration by iron or steel is often objectionable. During *synthesis* of phenol, as opposed to recovery of naturally occurring phenol, corrosion problems are complex, because of the use of mineral acids and alkalies. Above 150°C (300°F), steel is attacked by by-product acids and aldehydes.

Process

The predominant process for synthesis of phenol in the United States is the cumene-hydroperoxide route. The oxidation of toluene is a second process, via benzoic acid. Other sulphonation processes and benzene plants have shut down, but the Raschig process is still extant.

Cumene-hydroperoxide process (Fig. 53.1)

Benzene (C_6H_6) is alkylated with propylene to cumene $[C_6H_5CH(CH_3)_2]$, which is oxidized to the hydroperoxide. There are

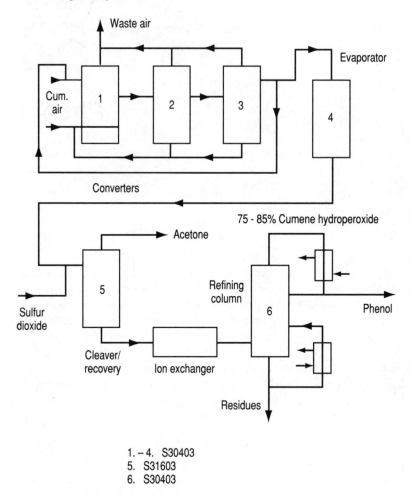

1. – 4. S30403
5. S31603
6. S30403

Figure 53.1 Cumene process.

two processes: (1) a low-pressure catalytic route (Allied/UOP) which predominates and (2) a high-pressure sodium-catalyzed process (Hercules-BP-Kellogg). The cumene is cleaved to produce a mole each of phenol and acetone.

Cumene is oxidized with air or oxygen at 80° to 130°C (175° to 265°F), sometimes with small amounts of caustic or sodium carbonate added as promoters and neutralizers. Typically, several oxidizer vessels are used in series, with 2 to 3% NaOH in the bottom of the first reactor, unreacted cumene being recycled. The cumene hydroperoxide is concentrated by evaporation and fed to the cleavage reactor, where it is carefully acidified with sulfur dioxide.

Some acetone is recovered overhead. The phenol-acetone crude product is ion-exchanged to remove acidity and distilled to produce the refined product.

Toluene process (Fig. 53.2)

Toluene is oxidized with air to benzoic acid at 140°C (285°F) over a cobalt catalyst, unreacted toluene being recovered overhead by distillation of the oxidation product. The benzoic acid is decarboxylated over a copper/magnesium catalyst at 240°C (465°F), losing one mol of carbon monoxide. Refined phenol is produced by distillation, produced water and unreacted benzoic acid being recirculated.

Sulphonation process (Fig. 53.3)

The sulphonation is effected sequentially, first in liquid-phase (benzene/9.5% oleum) and then in vapor-phase reaction of the presulphonated mixture. Wet benzene is recovered from the sulphonator as vapor, scrubbed free of acids, decanted, dried, and recirculated to the vapor-phase sulphonator. These preliminary operations are effected in steel and cast iron. The benzenesulphonic acid product from the base of the sulphonator is fused with concentrated caustic and made into a molten anhydrous sodium phenolate. The crude product is diluted with water and centrifuged to remove insoluble sodium sulfite, the phenolate solution feeding the lead-lined acidifier to spring phenol with dilute sulfuric acid and precipitate sodium sulfate. Crude, wet phenol is treated with concentrated sulfuric

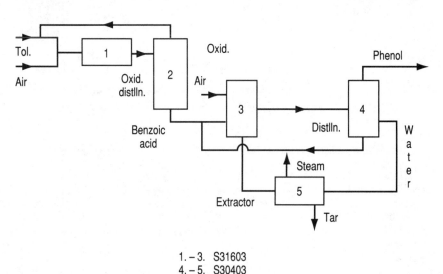

1. – 3. S31603
4. – 5. S30403

Figure 53.2   Toluene process.

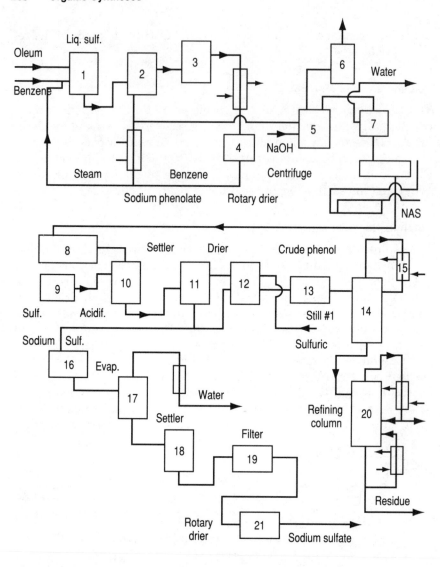

1. – 2. Ductile iron (F10001)
3. – 4. Carbon steel
5.   30 Ni Cast Iron (F41004)
6. – 9. Carbon steel
10. Carbon steel with 12 Cr (S41000) internals
11. Carbon steel

12. Carbon steel with S41000 internals
13. – 14. Carbon steel
15. Nickel N02200
16. – 18. Carbon steel
19. Carbon steel with Ni-Cu (N04400) internals
20. Nickel N02200
21. Carbon steel

Figure 53.3 Sulphonation process.

acid in a lead-lined drier, the dilute acid being recycled; the crude phenol is refined in a distillation system. The final still and condenser utilize nickel alloy 200 to retain product purity.

Materials

Aluminum and its alloys

Phenol is compatible with aluminum below 150°C (300°F), traces of water (0.3% minimum) acting as an inhibitor. Aqueous solutions have little action up to at least 100°C (212°F).

Steel and cast iron

Where iron contamination and discoloration is not objectionable, steel and cast iron can be employed. Steel may be used for crude product storage and for still bottoms piping in some instances.

Stainless steels

Conventional austenitic grades are compatible with phenol (e.g., S30403). High-alloy grades (e.g., CN7M) have been used for valves and pumps, particularly in the sulphonation process. Type 316L (S31603) is the preferred material where erosion-corrosion is encountered with carbon steel and for handling "phenolic water" (i.e., 80/20 or 90/10 water/phenol solutions), which is excessively corrosive to steel and low-alloy steels.

Copper and its alloys

Copper and arsenical yellow brasses are not suitable for phenol, particularly if traces of oxygen or oxidants are present. Chromophoric corrosion products are objectionable.

Nickel alloys

Nickel 200 (N02200) has been widely used for columns, exchangers, and piping systems for high-purity phenol, while nickel-clad steel tanks have been employed for storage. Alloy 600 (N06600) was used in the caustic fusion reaction with benzene sulphonic acid.

Reactive metals

Tantalum bayonet heaters are used for HCl vapor recovery, but the reactive metals have no direct application in phenol service.

Precious metals

Silver, gold, and platinum are not utilized for phenol service, since less expensive alloys are suitable where needed.

Other metals and alloys

Lead is moderately resistant [less than 0.5 mm/a (20 mpy)]. Tin is approved for shipping, but zinc is not.

Nonmetallic materials

Organic. The fluoroplastics (e.g., FEP, PTFE, PFA) are resistant to about 180°C (360°F), but PVDF is limited to about 66°C in strong phenol and 100°C (212°F) in 10% phenol. Standard plastics, including the fiberglass-reinforced materials, show quite poor resistance. Furanes and phenolics will tolerate 5 to 10% solutions to about 100°C, but not strong phenol. Polyethylene (other than HDPE) is substantially nonresistant, and polypropylene is limited to 5 to 10% solutions at room temperature.

Silicone rubber is reportedly resistant to about 100°C (212°F), and butyl rubber to 80% phenol at room temperature. Otherwise, rubber and elastomers are nonresistant to phenol solutions.

Wooden tanks have been used to handle dilute carbolic acid solutions to about 80°C (175°F).

Carbon and graphite equipment is inherently resistant, but the manufacturer should be consulted regarding suitability of any cemented joints or fillers. Probably only carbon-filled furane cements are trustworthy.

Baked phenolic coatings show good resistance but are not usually employed because of iron pickup at holidays in the coating.

Inorganic. Glassware and brick-lined equipment are used as required in certain parts of phenol production, but otherwise are not required for phenol itself except for carboy shipment.

Pitfalls

Aluminum will react rapidly with phenol above 150°C (300°F), sometimes after a brief incubation period. The attendant evolution of hydrogen poses a danger of fire and explosion. More than 0.3% water will inhibit the attack, but is not reliable at high temperatures.

Although steel is resistant up to about 205°C (400°F), the rate of attack is increased severalfold in the presence of sulfur contamination.

Sulfur contamination will adversely affect nickel-base alloys at elevated temperature.

Storage and Handling

Phenol is toxic if absorbed through the skin and can cause painful burns. It is classified as a class B poison by DOT regulations.

Pure phenol is a white crystalline solid, sold as 98% purity minimum. It may be shipped in heated containers above 41°C (106°F). Corrosion-resistant materials are used to prevent formation of chromophores which discolor the product. Use of the following materials of construction is common for handling and storage and is thought to constitute good engineering practice:

Drums	S30403; aluminum (with .3% water)
Tanks	S30403 or S31603; glass-lined steel
Tank trucks	S30403 or S31603
Railroad cars	S30403 or S31603
Piping	S31603 (traced)
Valves	CN7M or CF3M
Pumps	CN7M or CF3M
Gaskets	PTFE/304 spiral-wound, graphite

References

1. G. E. Moller, "Corrosion by Phenol," private communication.
2. J. E. Lee, *Materials of Construction for Chemical Process Industries,* McGraw-Hill, New York, 1950.

Styrene

Introduction

Styrene (vinyl benzene), an unsaturated aromatic monomer, is used in the manufacture of plastics and elastomers (e.g., ABS, styrene-butadiene rubber [SBR]).

Process

The greatest volume of styrene is produced by the catalytic dehydrogenation of ethyl benzene (Chap. 44) in the presence of steam (Fig. 54.1). The catalysts are based on ferric oxide. The crude styrene is purified by vacuum distillation in the presence of inhibitors to minimize polymerization (e.g., nitrophenols).

Materials

Aluminum alloys

Aluminum and its alloys are resistant to styrene. Polymerization of the monomer is effected in aluminum to prevent product discoloration.[1]

Steel and cast iron

Styrene is noncorrosive to steel and cast iron, and steel is approved for shipment of the monomer.

Stainless steels

Any grade of stainless steel is compatible with styrene.

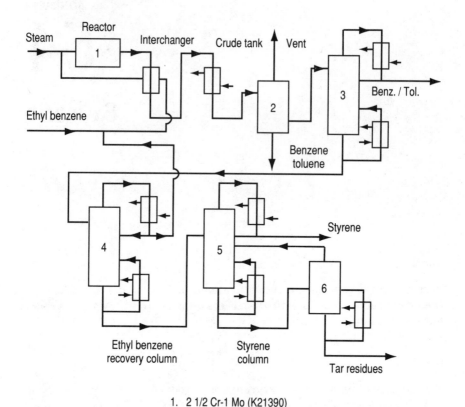

1. 2 1/2 Cr-1 Mo (K21390)
2. – 6. Carbon steel

Figure 54.1   Styrene (ethyl benzene process).

Copper and its alloys

Copper and its alloys resist styrene but are not recommended for storage or shipment of the monomer because of chromophoric compounds. The dimer is stable in copper.[2]

Nickel alloys

All nickel-base alloys are compatible with styrene, but are not normally used because of the suitability of conventional stainless steels.

Reactive metals

Titanium, zirconium, and tantalum are fully resistant, but there is no occasion to require them.

Precious metals

Silver, gold, and platinum find no application in this service.

Other metals and alloys

Lead, zinc, and tin are fully resistant. The latter two are approved for shipping containers.

Nonmetallic materials

Organic. Because of the aromatic solvent nature of styrene, thermoplastic materials are unsuitable except the fluorinated varieties. The polyamides are resistant, while in the glass-reinforced thermosets only the epoxies are unreservedly suitable. In the polyester formulations, some are deemed suitable; consult the manufacturer for specific applications.

Certain epoxy-phenolic and some urethane coatings are approved for this service.

Of the synthetic rubbers and elastomers, only the PVDF-hexafluoropropylene polymer (Viton) is suitable even at ambient temperature.

Inorganic. All glassware and ceramicware is suitable, but is not required for this service.

Pitfalls

There are no reported pitfalls in handling the monomer. Contaminants could cause polymerization under some conditions.

Storage and Handling

Use of the following materials of construction is common for handling and storage and is thought to constitute good engineering practice:

Tanks	Steel
Tank trucks	Steel
Railroad cars	Steel
Piping	Steel
Valves	Steel (S41000 trim)
Pumps	CF8M or CF8
Gaskets	Spiral-wound PTFE/stainless steel

References

1. I. Mellan, *Corrosion Resistant Materials Handbook,* 3d ed., Noyes Data Corp., Park Ridge, N.J., 1976.
2. Private communication.

55

Urea

Introduction

Urea is an organic product (NH_2CONH_2), synthesized from inorganic reactants (ammonia and carbon dioxide), used in fertilizer as a convenient form for fixed nitrogen and in process applications (e.g., urea-formaldehyde resins). It is an intermediate in manufacture of sulfamic acid and related compounds. Urea is a crystalline material, m.p. 132.7°C (270°F), but is also sold as a liquid grade, although in lesser quantities.

Process (Fig. 55.1)

The process involves reaction of 2 mol ammonia with 1 mol carbon dioxide at about 20 MPa (3000 lb/in^2) and 190°C (375°F). A corrosive intermediate, ammonium carbamate, is responsible for the need for corrosion-resistant construction:

$$2NH_3 + CO_2 \rightarrow NH_2COONH_4 \rightarrow NH_2CONH_2 + H_2O$$

A patented process involves the use of air or oxygen injection in an attempt to maintain passivity of type 316L construction. Currently, more highly alloyed stainless steels (25 Cr-22 Ni-2 Mo; S31050) are favored. In other processes, titanium linings (in multilayered steel vessels) or claddings are used to combat the carbamate corrosion, and one producer has had to go to zirconium linings.

From the autoclave, the melt is cooled to about 150°C (300°F) and fed to a urea still operating at about 60°C (140°F). Unconverted ammonia and carbon dioxide are recovered overhead, while a water solution of urea is delivered to a crystallizer operating at about 15°C (60°F). More ammonia is recovered, while the magma feeds a contin-

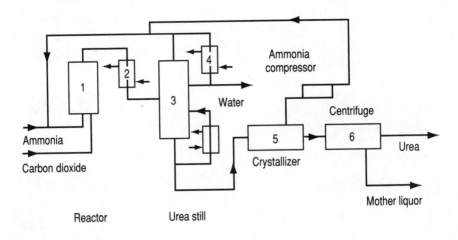

1. – 2. 30 Cr-25 Ni (S31003)
3. – 6. S31603

Figure 55.1   Urea synthesis.

uous crystallizer to produce the crystalline product. The mother liquor
can be utilized as liquid fertilizer.

Materials

Aluminum and its alloys

Urea and urea solutions may be handled in aluminum at ambient
temperatures.[1] At mildly elevated temperatures, anodizing is beneficial. Aluminum process vessels, tanks, and piping are conventional.

Steel and cast iron

Corrosion of steel is reportedly less than 0.5 mm/a (20 mpy) in dilute
solutions at ambient temperature.[2] Higher rates are observed at
higher concentrations and temperatures.

Stainless steels

Rates for conventional austenitic grades are less than .25 mm/a (1
mpy) to about 120°C (250°F), above which temperature rates can es-
calate rapidly.[3] Passivity is lost unless oxidizing conditions are maintained. The superferritic alloy S44626 reportedly has excellent resistance compared to S31603.[4]

Copper and its alloys

Copper and its alloys are not used in this service because of the ammoniacal nature of the process.

Nickel alloys

Nickel-base alloys offer no advantage in this application and are not used in practice, since other, less-expensive materials are suitable.

Reactive metals

Titanium is highly resistant, with not more than 0.1 mm/a (3 mpy) at 180°C (360°F). Zirconium has been used under more corrosive conditions in the process.

Precious metals

Silver is reported to have excellent resistance and has been used in the process previously.[2]

Other metals and alloys

At one time, lead-lined equipment was used in the urea process up to about 190°C (375°F).[1] Even magnesium is reported resistant to water solutions, at ambient temperature only.[4]

Nonmetallic materials

Organic. Conventional thermoplastic and thermosetting formulations may be used within normal temperature limits.[4]

Conventional rubbers and elastomers, with the possible exception of Buna N, are useful to about 75°C (170°F).

Silica, furane, and sulfur cements resist urea solutions at elevated temperatures. Phenolic resin cements are not resistant.

Inorganic. Glassware and ceramicware are resistant to about 100°C (212°F).

Pitfalls

Steel vessels may suffer corrosion due to ingress of corrosive carbamates at defects in the lining, followed by buckling of the lining.

Any kind of stainless steel in this process is dependent on the maintenance of oxidizing conditions. The process is highly conducive to intergranular corrosion if carbon contamination of low-carbon grades occurs.

Titanium linings may suffer corrosion, embrittlement, and/or cracking due to hydrogen pickup from iron contamination of the titanium plate or welds.[5] The iron contamination can come from use of steel tools or wire brushes on the titanium or be left from fabrication practices if the titanium was not chemically cleaned or anodized. Titanium welds may be deficient in resistance if improperly shielded during welding.

Storage and Handling

Use of the following materials of construction is common for handling and storage and is thought to constitute good engineering practice:

Tanks	Aluminum or S30400
Tank trucks	Aluminum or S30400
Railroad cars	Aluminum or S30400
Piping	Aluminum or S30403
Valves	CF8M
Pumps	CF8M or CN7M
Gaskets	Neoprene, graphite fiber, spiral-wound PTFE/stainless steel

References

1. I. Mellan, *Corrosion Resistant Materials Handbook*, 3d ed., Noyes Data Corp., Park Ridge, N.J., 1976.
2. N. E. Hamner, *Corrosion Data Survey*, 5th ed., NACE, Houston, 1974.
3. J. D. Polar, "A Guide to Corrosion Resistance," Climax Molybdenum Co., New York.
4. B. D. Craig, *Handbook of Corrosion Data*, ASM International, Metals Park, Ohio, 1989.
5. P. E. Krystow, "Materials and Corrosion Problems in Urea Plants," *Chemical Engineering Progress*, vol. 67, no. 4, April 1971.

Vinyl Acetate

Introduction

Vinyl acetate is a colorless flammable liquid whose pleasant odor quickly becomes irritating. It is an important precursor for polyvinyl acetate (PVA) and its copolymers (used in paints, adhesives, and plastics), as well as for polyvinyl alcohol.

Process

The original process of reaction between acetylene and acetic acid has been largely replaced by the palladium-catalyzed, vapor-phase reaction of ethylene and acetic acid at about 175° to 200°C and 0.4 to 1.0 MPa (58 to 145 lb/in^2) (Fig. 56.1). Inhibitors, such as diphenylamine or hydroquinone, are added to prevent polymerization.[1]

In the Wacker process, ethylene is reacted in anhydrous acetic acid, followed by oxidation of Pd by $CuCl_2$ and of the resulting CuCl by oxygen.

Materials

Aluminum alloys

Aluminum is resistant to vinyl acetate and is widely used to handle the finished product.[2,3] Unaccountably, high rates are reported in one source, probably due to minor dilution or metal ion contamination.

Steel and cast iron

Steel and cast irons are resistant to this product, and steel is approved for shipment of the monomer.[4]

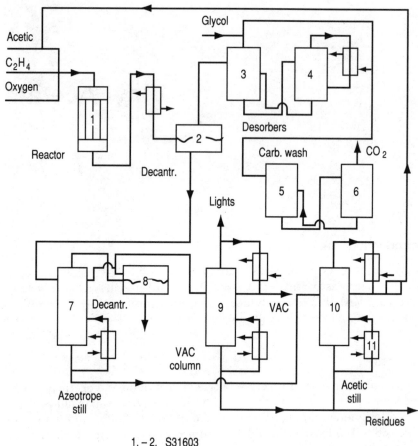

1. – 2. S31603
3. – 6. S30403
7. – 10. S31603
11. Ni-Cr-Mo (N10276) or impervious graphite

Figure 56.1 Vinyl acetate (vapor phase process).

Stainless steels

Any grade of stainless steel is resistant to the monomer. S31603 is preferred at moderately elevated temperature, e.g., less than 70°C (160°F).

Copper and its alloys

Copper and its alloys are not approved for shipment or storage because of chromophores. An earlier objection to copper-base materials was concern over acetylide formation in the older process; it is no longer a valid concern.

Nickel alloys

Nickel-base alloys are not required for this service.

Reactive metals

The reactive metals are fully resistant to vinyl acetate.

Precious metals

Silver, gold, and platinum are unaffected, but find no application in this service.

Other metals and alloys

Lead would be unsuitable if even traces of acetic acid and air are present, but tin and zinc are approved for shipment and storage.[3]

Nonmetallic materials

Organic.   The ester characteristics and vinyl group limit applications of some plastic formulations. Vinyl formulations are not resistant. Polyethylene and polypropylene are reported suitable by one source up to about 60°C (140°F).[5] However, they are not recommended for storage.[1,2] Chlorsulfonated polyethylene is resistant to about 60°C (140°F). Fluorinated plastics are fully resistant to their normal temperature limits.[5] Epoxy FRP, but not polyester, is suitable at ambient temperature.[1]

Among the elastomers, probably only the PVDF/hexafluoropropylene is suitably resistant.

Inorganic.   Glassware and ceramicware are resistant, but esters tend to dissolve cement or concrete even when acid-free.

Pitfalls

There are no reported pitfalls associated with the processing or handling of vinyl acetate.

Waste streams from the *polymerization* of vinyl acetate may contain organic peroxides, which are highly corrosive to almost all common materials of construction (including conventional stainless steels in the simultaneous presence of chlorides).

Storage and Handling

Use of the following materials of construction is common for handling and storage and is thought to constitute good engineering practice:

Tanks	Steel,* aluminum, S30403
Tank trucks	Aluminum, S30400
Railroad cars	Aluminum, S30400
Piping	S30403
Valves	CF8M
Pumps	CF8M
Gaskets	Graphite fiber, spiral-wound PTFE/stainless steel

References

1. G. T. Austin, *Shreve's Chemical Process Industries,* 5th ed., McGraw-Hill, New York, 1984.
2. I. Mellan, *Corrosion Resistant Materials Handbook,* 3d ed., Noyes Data Corp., Park Ridge, N.J., 1976.
3. Private communication.
4. N. E. Hamner, *Corrosion Data Survey,* 5th ed., NACE, Houston, 1974.
5. P. A. Schweitzer, *Corrosion Resistance Tables,* Marcel Dekker, New York, 1986.

---

*If iron contamination is not a concern.

Vinyl Chloride and PVC

Introduction

Vinyl chloride monomer (VCM), a colorless, odorless gas at normal temperatures and pressures, is a high-volume commodity used as a precursor to polyvinyl chloride (PVC) and other thermoplastic copolymers. Under U.S. Occupational Safety and Health Administration (OSHA) regulations, it is usually handled as a liquid (b.p. −13.4°C), with no human contact.

Process

The original process of reaction between acetylene and hydrogen chloride has been replaced with methods based on ethylene dichloride [1,2-dichloroethane (EDC)].

In practice, EDC is produced from chlorination of ethylene and pyrolyzed at about 550°C (1020°F) and 3 MPa (618 lb/in^2) to form VCM and HCl (Fig. 57.1). The pyrolysis furnace has been constructed of both alloy 600 (N06600) and of heat-resisting castings, such as HD-30 (J93005) or HK-40 (J94204). The HCl is recirculated through an oxychlorination reaction, in which ethylene, oxygen, and HCl produce more EDC for pyrolysis. Corrosion-resistant materials are required to resist traces of HCl.

PVC is produced by a relatively simple polymerization of the monomer in type 304L (S30403) vessels, the resin being dried and packaged (Fig. 57.2). In recent years, environmental concerns have required recirculating waste process water, and the resultant lower pH and higher chloride content of the reaction mixture has required higher alloys to resist pitting and SCC.

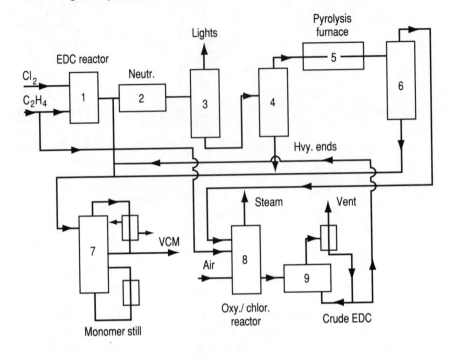

1. Brick-lined steel
2. Ni-Cr-Mo (N10276)
3. Ni-Cu (N04400)
4. Carbon steel
5. Ni-Cr-Fe (N06600)
6. Ni-Cu (N04400)
7. 20 Cr-25 Ni-6 Mo (S31254)
8. Brick-lined steel
9. Ni-Cr-Mo (N10276)

Figure 57.1   Vinyl chloride (EDC process).

Materials

Aluminum alloys

Aluminum and its alloys must be precluded from vinyl chloride service, since violent reactions may occur.[1]

Steel and cast iron

Steel and cast iron are resistant to *anhydrous* VCM, but even traces of water cause serious attack. Steel is approved for shipment and storage.[2]

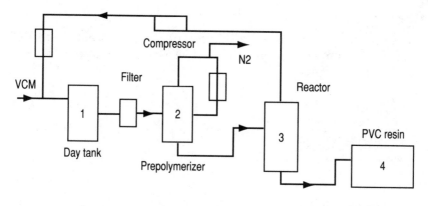

1. – 4.   S30403

Figure 57.2   PVC process.

Stainless steels

The conventional stainless steels, e.g., S30403 and S31603, are fully resistant to VCM itself. The latter molybdenum-bearing grade gives some advantage in the event of potential water contamination. See "Pitfalls" below. The high-performance 6% molybdenum grades such as S31254 or N08060 are preferable in water solutions with prolonged residence time. The high-alloy castings, such as HK-40 (J94244), are used in EDC pyrolysis furnaces for VCM production.

Copper and its alloys

Copper and its alloys, other than high-zinc brasses, should be suitable in the monomer by itself. High corrosion rates sometimes reported probably reflect hydrochloric acid and air contamination.[3] Also, there is a long-standing tradition against exposure of copper-rich alloys because of the possibility of copper acetylide formation in the older acetylene-based processes.

Note: Even ethylene may contain traces of acetylene compounds.

Nickel alloys

All nickel-base alloys resist VCM itself and even traces of free HCl. The copper-bearing alloy 400 (N04400) would be the least reliable, because of possible accretion of cupric ions. Nevertheless, it is rated excellent in anhydrous VCM to about 230°C (450°F).[3] Alloy 600 (N06600) has been successfully used in EDC pyrolysis in lieu of HK-40.

Reactive metals

Titanium has no application in this service, being susceptible to HCl attack. Zirconium and tantalum should be excellent, but there is little incentive for their use.

Precious metals

Silver is fully resistant to VCM, but the precious metals find no application in this service.

Other metals and alloys

Lead is reportedly unsuitable for VCM exposure, with rates less than 1.2 mm/a (50 mpy).[4] This may be a result of hydrochloric acid attack, since another source shows rates of less than 0.1 mm/a (2 mpy) in anhydrous VCM at ambient temperature.[3] Tin is approved for shipment or storage, but zinc is not.

Nonmetallic materials

Organic. As a chlorinated solvent, VCM is incompatible with many thermoplastics. Polyethylene containers are not resistant.[2] The fluorinated plastics should be fully resistant. Nylon is reportedly resistant.[4]

Baked-phenolic coatings are approved for shipping and storage.[2]

Of the rubbers and elastomers, only VDF/hexafluoropropylene and a specially compounded modified olefin are reported suitable as polymeric hose materials.

Inorganic. Glassware and ceramicware would be fully resistant to VCM.

Pitfalls

The primary pitfalls associated with VCM are related to its possible solvent action (traces in hydrocloric acid will destroy the conventional rubber lining used for acid storage) and hydrolysis which liberates HCl, with the attendant corrosion problems.

Storage and Handling

VCM is a carcinogen, and every precaution must be taken to prevent human contact. PVC is a dry resin product prepared to be free of residual monomer.

Use of the following materials of construction is common for handling and storage and is thought to constitute good engineering practice:

Tanks	Steel or S30403
Tank trucks	Steel or S30403
Railroad cars	Steel or S30403
Piping	Steel or S30403
Valves	Steel (N04400 trim) or CF8M
Pumps	CF3M
Gaskets	Spiral-wound PTFE/stainless steel

References

1. C. P. Dillon, "Aluminum in Potentially Corrosive Organic Solvents," *Materials Performance,* vol. 29, no. 11, November 1990.
2. Private communication.
3. N. E. Hamner, *Corrosion Data Survey,* 5th ed., NACE, Houston, 1974.
4. I. Mellan, *Corrosion Resistant Materials Handbook,* 5th ed., Noyes Data Corp., Park Ridge, N.J., 1976.

Part

5

Special Topics

This section covers miscellaneous problems associated with plant utilities and discusses in more detail specific problems noted briefly in the prior sections. It is an attempt to promote practical information which may be difficult to find in the conventional literature.

For more detailed information, the reader should consult the references.

Cooling Water

Introduction

Even though materials selection for chemical processes has been carefully addressed, many corrosion problems will arise if potential cooling water effects are not considered. Corrosion of certain materials and scaling by calcareous deposits are primary problems, but secondary effects arise from concentration cells, localized enrichment of specific ions, and from direct and indirect effects of bacterial action.

The nature and extent of cooling water problems will vary with the nature of the system (i.e., whether once-through or recirculated) and the chemistry of the water. Once-through water, because of high-volume usage, cannot economically be inhibited or treated for corrosion and scale control. Consequently, one must rely on proper materials selection, which, in water-cooled heat exchangers, means selecting for *both* process and waterside conditions. In recirculated systems, there is relatively little water loss (blow-down) relative to the circulation rate, and chemical treatment against scale and corrosion is economically feasible.

Corrosion

Corrosion by water is entirely caused, at cooling water temperatures, by dissolved gases (i.e., oxygen, carbon dioxide) and salts (e.g., sodium and other chlorides, bicarbonates, sulfates). Water that is not prone to scaling is generally more aggressive than a "hard" water, as described further below, but it is important to note that harder waters are not necessarily noncorrosive.

The minimum information necessary to evaluate corrosion and scale behavior of cooling water is shown by a typical water analysis, which defines the following parameters:

Variable	Units
Total dissolved solids (TDS)	ppm
Calcium, as calcium carbonate	ppm
Magnesium, as calcium carbonate	ppm
Total hardness (TH = Ca + Mg) (as calcium carbonate)	ppm
Phenolphthalein alkalinity (PA) (as calcium carbonate)	ppm
Methyl orange alkalinity (MOA) (as calcium carbonate)	ppm
pH	
Chloride as Cl^-	ppm
Sulfate as SO_4^{--}	ppm
Silica as SiO_2	ppm
Iron as Fe	ppm
Dissolved oxygen (DO)	ppm

The expression of some components as calcium carbonate equivalents is to facilitate calculation of scaling tendencies, as described further below.

From a practical standpoint, the major component affecting corrosion by water is the chloride ion. Chlorides directly affect corrosion of iron and steel, with corrosion reaching a maximum at about 6000 ppm chloride concentration (about 1% NaCl). Above that concentration (e.g., in seawater), corrosion is offset by a diminished oxygen solubility. The combined effect of chlorides with velocity adversely affects not only steel but also copper-base alloys. At high salinities and with the normal velocities encountered in heat exchangers [i.e., 2.4 to 3.6 m/s (8 to 12 ft/s)], 90-10 cupronickel (C70600) is the preferred copper alloy. Chlorides are also responsible for localized corrosion (e.g., pitting of aluminum, stainless steels, and other chromium-bearing alloys) and stress-corrosion cracking of austenitic stainless steels. When water-side conditions are conducive to pitting and/or SCC, high-performance stainless steels (e.g., S31254, N08367, S31803) or nickel-rich alloys (N08028, N08320) must be used, even though conventional grades like S30400 or S31600 might be resistant to the process-side conditions.

A particularly insidious corrosion phenomenon is crevice corrosion, which may occur either in mechanical crevices (e.g., threaded connections, flange faces, rolled joints) or under scales, films, or deposits. Crevice corrosion occurs primarily because of oxygen concentration cells (although dissolved sulfur cells are also important), with the oxygen-deficient area functioning as the anode and suffering accelerated attack.

Chlorides exacerbate the problem because of a lowered pH in the anode area as well as the specific ion effects. Although research studies have quantitatively defined critical crevice dimensions vs. specific pH and chloride ranges, crevice corrosion is best combated by eliminating crevices where possible (e.g., by tube-to-tubesheet welding) and by selecting resistant materials.

Although cooling waters tend to be aerobic, because of surface water exposure to the atmosphere or cooling-tower effects, anaerobic conditions may arise under shutdown (stagnant) conditions or under scales or deposits. Under such conditions, sulfate-reducing bacteria may produce hydrogen sulfide and dissolved sulfur (DS), which can have corrosion effects on steels and on copper alloys. Some aerobic bacteria can also cause corrosion phenomena (see Chap. 63, "Microbiological Corrosion").

The corrosion resistance of specific materials in water is discussed further below.

Scaling

Scaling occurs primarily because of the inverse solubility of calcium and magnesium carbonates, which are less soluble at higher temperatures. Heat-transfer surfaces are most prone to scaling, and this of course diminishes effective heat transfer, besides providing opportunity for crevice corrosion and for concentration of chlorides by absorption in the calcareous deposits.

In most freshwater cooling systems, the tendency for scaling can be calculated by using the Langelier saturation index (LSI) and factors based on pH, pH_S, TDS, Ca, MOA, and temperature.[1] In recirculated systems, the water chemistry can be modified to prevent scaling. In once-through systems, it is sometimes possible to size the exchangers so that tube-wall temperatures do not exceed those which would cause scale deposition to occur.

Materials

Iron and steel

It should be assumed that cooling-water systems are corrosive to iron and steel unless specifically inhibited, which is feasible only in recirculated systems. Do *not* use carbon-steel piping or vessels in uninhibited cooling water.

Gray-cast-iron or ductile-iron valves and pumps may be used in water of potable quality, being perhaps 30 to 50% more resistant than steel and much more massive than wrought products. However, soft or acidic waters may cause graphitic corrosion. Graphitic corrosion products can cause galvanic corrosion of other metals in electrical contact with the cast-iron components.

Galvanized steel piping may be used in fresh waters of approximately potable quality (e.g., 250 ppm chloride maximum). The zinc layer is inherently more resistant than steel, acting as a temporary barrier coating. However, once the coating is penetrated by mechanical or corrosion damage, exposing the steel substrate, the zinc will suffer rapid galvanic corrosion to afford cathodic protection to the underlying steel.

Note: Galvanized pipe must always be electrically isolated from valves, pumps, etc. of other alloys to prevent premature wastage of the zinc. This is best accomplished by means of armored insulated bushings or couplings, although PTFE tape is sometimes used in simple threaded connections.

Aluminum alloys

Aluminum and its alloys are quite resistant to natural waters under *flowing* conditions. They are, however, prone to chloride pitting under deposits and to "cementation" corrosion if heavy metal ions are present in the water.

Copper and its alloys

Although copper and its alloys can be severely attacked by soft waters in the presence of oxygen and carbon dioxide, they are usually resistant to cooling waters of various chemistries. In the process industries, the specific alloys are chosen to resist general corrosion, dezincification and erosion-corrosion.

The transition from essentially pure copper to brasses and cupronickels is dictated by the combined effects of velocity and chloride content. In fresh waters, the first choice is Admiralty metal (inhibited with arsenic, antimony, or phosphorus) for heat-exchanger tubing to resist dezincification. In more saline waters, including seawater, an inhibited aluminum brass has been used. However, in modern usage, the preferred alloy for either high velocity or high salinity is the iron-bearing 90-10 cupronickel alloy (C70600).[2]

Stainless steels

Both 12% and 17% straight chromium grades of stainless steel find application in low-chloride waters where process conditions indicate their selection. The extra-low interstitial superferritic grades such as

26-1 (S44700) are much preferred in waters of chloride content above about 250 ppm.

The conventional austenitic stainless steels typified by S30400 and S31600 may be used in chloride-bearing waters to at least 1000 ppm, provided the tube-wall temperature is not above 50° to 60°C. Water velocity must be maintained above about 0.9 m/s (3 ft/s), and there must be no situation in which chlorides can concentrate. There are reports of SCC in vertical condensers with water on the shell side at Cl⁻ concentrations as low as 50 ppm. (Such exchangers require an adequate and operable tubesheet *vent.*)

If these conditions cannot be met, or if the chloride content is above about 1000 ppm, the high-performance stainless steels of about 6% molybdenum content should be selected. The duplex grades (e.g., S31500, S31803, S32550) are often chosen. However, to cope with aggressive chemical exposures simultaneously, the austenitic grades (S31254, N08367) are preferred, since they are less subject to attack by reducing acid conditions.

Nickel alloys

In aggressive waters, or as process conditions require, both nickel-rich alloys (e.g., N08825, N08028, N06007) and nickel-base alloys (e.g., N06625, N06022, N10276) may be employed.

Nickel alloys 200 (N02200) and 400 (N04400) may be employed as required, but waterside conditions may demand more resistant alloys. When process-side conditions call for nonchromium, high-molybdenum alloys (N10665), high-chloride waters may require the chromium-bearing high-molybdenum alloys.

In special circumstances in which no single metal or alloy will suffice, bimetallic tubes may be utilized (e.g., S31600/C70600) to resist both chemical corrosion and cooling-water attack.

Reactive metals

When titanium or zirconium is required for process conditions, it can usually be considered resistant to any cooling water under conventional conditions of operation. (Titanium has been known to pit in seawater under excessive temperature load or alternating wet-dry conditions.)

One must not use steel accoutrements with titanium tube bundles when water is on the shell side, since there will be severe corrosion of baffles and tie-rods, often followed by vibration and abrasive cutting of the tubes themselves. Stainless or high-copper alloys are suitable in most cases. With water on the tube side, the water boxes should be titanium or FRP to minimize galvanic corrosion on lower alloys by the titanium tube bundle.

Nonmetallic materials

Any thermoplastic and thermosetting resin products are useful in cooling waters within their normal operating temperature and pressure limitations. They are not subject to the vagaries of water chemistry, as are the metals and alloys.

Halocarbon and ceramic-type exchangers pose no problems in cooling-water systems, provided there is no through-permeation of process-side chemicals (e.g., hydrogen chloride).

In tube-and-shell impervious graphite exchangers, corrosion-resistant hardware must be used. The carbon-steel galvanic couple is unacceptable in cooling-water service.

References

1. C. P. Dillon, *Corrosion Control in the Chemical Process Industries*, McGraw-Hill, New York, 1986.
2. C. P. Dillon, "Seawater-Cooled Tube-and-Shell Heat Exchangers in the Chemical Process Industries," publication no. 26, Materials Technology Institute, St. Louis, 1984.

Wastewater

Introduction

Wastewaters from industrial processes may contain substantial amounts of inorganic or organic species which can profoundly affect corrosion characteristics. Oxidizing inorganic cations (e.g., Fe^{+++}, Cu^{++}) or anions (e.g., nitrates) or reducing species such as Sn^{++} or hydrogen sulfide may be present. Organic contaminants may render the system anaerobic by consumption of dissolved oxygen [measured as chemical oxygen demand (COD)] and can affect organic materials or coatings because of absorption mechanisms. Micro- or macro-organisms can create a biological oxygen demand (BOD). In addition to changes in the redox potential arising in this manner, there can be specific ion effects, as from chlorides, fluorides, sulfates, sulfides, etc.

Corrosion

Corrosion by wastewater is caused by dissolved gases (i.e., oxygen, carbon dioxide, oxides of sulfur or nitrogen, hydrogen sulfide), various salts, and organic contaminants. The pH or total acidity (which is more important than pH where weak acids or acid salts are concerned) may be radically different than that encountered in cooling water.

The usual analytical data described for cooling water in the previous chapter should be supplemented by data on known or suspected contaminants. Oxidizing and reducing species are of particular interest, and chloride content is as important as in cooling waters. Organic chlorides are an important source of further chloride contamination by hydrolysis or pyrolysis (see Chap. 60) and pose a solvent-attack hazard to plastic and elastomeric materials. Sulfates are inimical to cement and concrete. Otherwise noncorrosive organic species (e.g., aromatic solvents) pose a solvent-attack hazard to rubber linings.

Domestic and municipal wastewaters tend to be anaerobic because of BOD, but process industry wastes vary from powerfully oxidizing (e.g., from nitrates or peroxides) to reducing in nature.

While it is difficult to be specific, because of the varieties of contaminants encountered, the corrosion resistance of the common materials in wastewaters is discussed further below.

Scaling

Scaling occurs primarily because of the inverse solubility of certain salts. In addition to the calcium and magnesium carbonates encountered in natural waters, wastewaters may contain other sparingly soluble compounds (e.g., sulfates, phosphates, silicates). Such problems must be addressed on an individual basis, since the LSI will not be applicable.

Materials

Iron and steel

Unlike cooling waters, which are assumed to be corrosive to iron and steel, wastewaters *may* be harmless in the neutral pH range (e.g., pH 6 to 9) if anaerobic and free from bacterial action. Cast-iron sewer pipe is usually employed with a cementitious lining which enhances its resistance in the absence of contaminants corrosive to portland cement.

Gray-cast-iron or ductile-iron valves and pumps may be used in wastewater of suitable chemistry.

The major caveats relative to iron and steel in wastewater pertain to acidic conditions and microbiological effects. At pH's below 5 with strong acids, or below about 5.5 with weak acids (e.g., carbonic acid, organic acids, acidic salts), acidic corrosion will occur with both iron and steel. Under anaerobic conditions, even at neutral pH's, bacterial action by sulfate-reducing bacteria can produce hydrogen sulfide, with its attendant problems.

Galvanized steel piping may be used in wastewaters known to be only mildly corrosive to steel, with a moderate improvement in service life. The usual precautions against galvanic contacts should be employed.

Aluminum alloys

Aluminum and its alloys should not be employed in wastewaters, as a rule, because of the many possible adverse contaminants. Exceptions may be made where the waste streams are specifically identified as noncorrosive to this group of materials.

Copper and its alloys

Copper and its alloys can be severely attacked by soft wastewaters in the presence of oxygen and carbon dioxide. They are corroded by many sulfur compounds and by ammoniated or amine compounds in the presence of oxidizing species. Ammonia is also specific for SCC of high-strength copper and copper alloys, except for cupronickel.

Stainless steels

Although straight chromium grades of stainless steel can be used in wastewaters, it is inadvisable to do so if there is any possibility of acid chlorides, sulfides, or other adverse contaminants being present.

The conventional austenitic stainless steels typified by S30400 and S31600 are widely used in wastewater systems with great success, provided only nominal amounts of chlorides are present. Otherwise, pitting, crevice corrosion and/or SCC may ensue, in which case high-performance grades (e.g., S31254, N08367) are indicated.

Nickel alloys

In aggressive wastewaters, both nickel-rich alloys (e.g., N08825, N08028, N06007) and nickel-base alloys (e.g., N06625, N06022, N10276) have been employed for critical components or for greatly extended life. Their use is primarily confined to threaded fasteners and selected slender structural components.

Nickel alloys 200 (N02200) and 400 (N04400) may be employed as required, in the absence of specific corrosive species, but are usually not economical.

Reactive metals

Titanium, zirconium, and tantalum are not usually specified for wastewater services because of their high cost and the reliability of other materials. Titanium would perhaps be useful in high-chloride wastes.

Nonmetallic materials

Thermoplastics, thermosetting resins, and elastomers may be used in wastewaters as they are with many corrosive waters. The important thing to remember is that the plastic or elastomer selected must be resistant to a *100% concentration* of any known organic contaminants. Otherwise, failure may occur due to selective absorption over an extended period of time (e.g., polypropylene vs. sorbic acid, FRP vs. chlorinated solvents or carbon disulfide, rubber vs. toluene or ethylene dichloride).

Cement-lined and concrete sewer lines have been commonly employed in municipal waste waters. Major problems arise from vapor-phase corrosion due to formation of sulfuric acid by thiobacillus bacteria under some circumstances. Any low-pH condition or calcium-deficient water will attack the portland cements, while high sulfate concentrations, even under neutral or alkaline conditions, can be harmful.

60

Environmental Cracking

Introduction

Environmental cracking is the brittle failure, usually of an otherwise ductile material, by cracking in the simultaneous presence of tensile stress and a specific corrodent. It manifests itself in three ways:

- Anodic stress-corrosion cracking (SSC)
- Cathodic hydrogen-assisted cracking (HAC)
- Liquid-metal cracking (LMC)

The *specificity* of the chemical environment for certain types of materials distinguishes this type of cracking phenomenon from corrosion fatigue.[1]

Environmental cracking is a particularly insidious form of localized corrosion in that it can be present at a microscopic level, initially undetectable by the naked eye, while seriously affecting the mechanical integrity of equipment.

Stress-Corrosion Cracking

SCC of metals is an anodic electrochemical process and therefore amenable to mitigation by cathodic protection. Application of anodic metals by hot dipping or metallizing has been used to delay SCC of stainless steels in brackish waters and of carbon steel in liquid ammonia.

Note: Environmental cracking of plastics, such as polycarbonates and polyolefins, is purely chemical in nature, and is called *chemical stress cracking* (ChSC).

In most situations, SCC occurs under *mildly* corrosive conditions, because more aggressive corrosion may preclude the development of the necessary anode-cathode relationship. A highly stressed area, although not necessarily the point of maximum stress, becomes anodic to the adjacent metal. The large cathode/anode area ratio then causes rapid penetration along grain boundaries or slip planes while the tensile stresses tend to pull the metal apart.

There is usually a relatively long initiation period, varying from hours to years, depending on the material and environment combination, followed by rapid cracking.

For some combinations (e.g., steel in hot caustic), there may be a threshold stress below which SCC will not occur. In others (e.g., 18-8 stainless steels in chloride environments), the threshold is absent or so low as to be of academic interest only.

The stresses must be tensile in nature, but may be either residual or applied. The former class comprises stresses remaining from contraction of weldments during cooling under their inherent restraint or from bending or forming during fabrication. Applied stresses derive from mechanical loading, pressurizing, and thermal expansion.

Any measure which reduces residual stress is beneficial, although the effect is finite. Thermal stress-relief is a practical measure, often doubling or trebling the time to failure. Note that thermal stress relief is most effective for tanks or vessels if they can be subjected to the necessary prolonged heating [usually about 2 hours per inch of thickness at the appropriate temperature; e.g., 600°C (1100°F) for steel, 1000°C (1750°F) for 18-8 stainless steel] without damage. Heat exchangers or tube bundles in the chemical process industries (CPI) should be stress-relieved only *in toto* after tubes have been strength-welded to tubesheet. (Stress-relieving tube U bends only prevents localized SCC but leaves the rolled and/or welded joints still susceptible.)

Mechanical stress relief can be effected by shot peening, which puts the metal surface into *compressive* stress. As long as the compressed surface is not removed by general corrosion, this approach is useful, particularly when the equipment does not lend itself to thermal stress relief (e.g., because of size or potential distortion).

If the agent responsible for SCC is present in only small concentrations, it may be selectively removed or neutralized. In a few cases, inhibition is feasible (e.g., by water in liquid ammonia). Where this is impractical or where the massive environment is specific, selection of SCC-resistant materials (e.g., cupronickel in lieu of brass, alloy 600 in lieu of steel, S31254 in lieu of S31603) or use of *heavy* barrier coatings or linings (to isolate the material from the environment) is required.

SCC is usually aggravated by elevated temperatures at least for as long as the environment remains specific for this mode of failure (e.g., aqueous chlorides).

In the CPI, the most common SCC failures relate to those involving the austenitic 18-8 stainless steels and the specific species responsible, especially as to critical concentrations of such. The chloride ion is both ubiquitous, being present in atmospheres, water, and processes, and highly specific for SCC of the workhorse alloys type 304 and type 316 and their low-carbon analogs.

Chloride concentration limits

The most commonly asked question relates to an acceptable concentration of inorganic chlorides, at and below which SCC will not occur. There is *no* answer for this question as posed, but we can give certain qualified answers.

Types of chlorides. The *most* common chloride is sodium chloride, as found in natural waters (NaCl), followed by calcium and magnesium chlorides in the same environment. The last, especially, is highly specific, as evidenced by the common laboratory test in boiling 42% $MgCl_2$, which will crack 18-8 alloy within a few hours.

The chloride ion in aqueous systems, derived from sodium chloride primarily, tends to be present in an essentially neutral and nonoxidizing situation. *If it can concentrate* (as in vapor spaces or splash zones), even 1 ppm is too much, because the concentration increases tremendously over a period of time. However, because of its solubility, it is relatively harmless in *total immersion* and *flowing conditions* below about 50° to 60°C (120° to 140°F) in fresh water (e.g., up to at least 1000 ppm) and even in seawater, so far as SCC (*not* pitting) is concerned. The proper approach in cooling water systems is to avoid crevices and vapor spaces, prevent scaling or deposits (see below), provide an operable vent for tubesheets in vertical condensers (when water is on the shell side), and design and operate the exchanger so that *tube-wall* temperatures do not exceed 50° to 60°C.

Any kind of film of foreign materials (e.g., silica, alumina, calcareous deposits) can adsorb, occlude, and otherwise concentrate chlorides to an unacceptable level, besides raising the metal temperature by interfering with heat transfer.

Oxidizing chlorides like ferric chloride ($FeCl_3$) increase the likelihood of SCC by shifting the corrosion potential into a critical area. *Rust films* deposited from waters (or even powdered iron or rust in the atmosphere) greatly increase the risk of SCC at otherwise innocuous chloride levels. The presence of specific oxidizing or reducing agents may negate conventional wisdom as regards aqueous chloride SCC.

Lastly, hydrogen chloride itself has been observed to cause SCC at temperatures of *minus* 20°F! *Any* acid pH below the nominally neutral range of water (e.g., about 5.5 to 8.5) is a potential problem.

We would warn the reader *particularly* not to transpose extraneous data to a given system! A paper by no less a personage than M. G. Fontana, professor emeritus at Ohio State, once described a municipal wastewater treatment system in which SCC of Type 304L did not occur below 200 ppm. This does *not* mean that this level applies in other systems and indeed the pH, contaminants, and redox potential relevant to the system studied were not defined—it was stated only that SCC did not occur below 200 ppm chloride in that particular system during the test period.[2]

Hydrolyzable/pyrolyzable chlorides. A distinction must be made between chlorine-containing compounds which hydrolyze to form inorganic chloride ions and those organic chlorocarbons which can only pyrolyze. Obviously, either process can form chloride ion contamination in a process, but the distinction is very important.

In the petroleum industry, the term *hydrolyzable chlorides* is usually applied to calcium and magnesium chlorides entering with the crude oil. At high temperatures, these do hydrolyze, releasing volatile HCl to travel overhead in the still system and leaving insoluble calcium and magnesium salts behind.

Hydrolyzable *organic* chlorides are another matter. For example, the asymmetrical ethylidene chloride is more readily hydrolyzed in hot water than the symmetrical isomer (ethylene dichloride). Vinyl chloride monomer is unsaturated and can only hydrolyze through an epoxy linkage, so its formation of HCl is controlled by the amount of dissolved oxygen in a water solution. On the other hand, the saturated low-molecular-weight polymer PVC, if returned in a waste stream recycle, hydrolyzes directly in *warm* water and immediately lowers the reactor pH while increasing chloride concentration.

Aromatic chlorocarbons with the chlorine atoms directly attached to the benzene ring (e.g., monochlorbenzene) are highly resistant to aqueous hydrolysis, but do *pyrolyze* at elevated temperatures.

Some published literature is misleading because the release of chlorides has been determined by pyrolyzing a suspected material in a Parr bomb and analyzing the water rinse for inorganic chlorides. This gives an erroneously high answer if the source is only pyrolyzible. From a practical standpoint, suspected organic materials (e.g., latex feed stocks) should be heated in *distilled water* for a few hours, after which pH and chloride concentration can be measured.

It is true, of course, that a great many organic oils, greases, paints, marking compounds and wrapping tapes have been properly identified as a source of inorganic chlorides responsible for SCC.

Catalytic effects must not be overlooked. For example, wet carbon tetrachloride is very resistant to hydrolysis in glassware. The presence of reagent-grade ferric oxide or of bare steel or mill scale (Fe_3O_4) has no effect. However, for some unknown reason, wet carbon tetrachloride *will* hydrolyze in the presence of *rusty* steel, liberating phosgene ($COCl_2$) and traces of HCl. We assume that this is caused by some obscure catalytic or electrochemical effect.

Chlorides vs. caustic. Hot concentrated sodium hydroxide is also specific for SCC of 18-8 stainless steels,[3] although it can also simply depassivate the alloy and cause rapid general corrosion.[4] Unfortunately, on microscopic examination, SCC by caustic has *exactly* the same morphology as chloride SCC in an annealed structure (although a sensitized structure will show intergranular, rather than transcrystalline, cracking).

The problem of identifying one causative agent may be critical (e.g., in steam contaminated with *both* chlorides and alkaline contaminants), as the decision profoundly affects selection of alternative alloys. Fortunately, SCC by caustic at temperatures of the order of 300°C (570°F) is accompanied by a "blueing" effect, which is a dead giveaway. If contamination cannot be eliminated, the proper replacement for 18-8 is alloy 625 or C276/C22. In the absence of blueing, an alloy 800 or 825 will resist SCC by chlorides at lower temperatures, although they are no better than 18-8 against caustic cracking.

Other alloys

Certain nickel-base alloys (such as N06600, N06625 and N10276) which are substantially immune to ordinary chloride SCC can suffer IGSCC in high temperature chlorinations at about 400° to 450°C (750° to 840°F). Thermal stress relief appears to be beneficial to alloy 600, which is more susceptible in cold-worked areas, but less so for the molybdenum-bearing grades, apparently due to age-hardening effects.

Hydrogen-Assisted Cracking

HAC is a *cathodic* process and is therefore aggravated by CP or by any circumstances which promote the entry of atomic hydrogen into the metal, e.g., electroplating, galvanic couples, or chemicals (S, As, Sb) which slow down the dimerization of nascent hydrogen.

In HAC, atomic hydrogen from the electrochemical corrosion process penetrates the metal lattice to weaken the bond at areas of triaxial stress. [In normal corrosion processes, the nascent atomic hydrogen generated by reduction of hydrogen ions (protons) either dimerizes for evolution as molecular hydrogen or is oxidized to water or hydroxyl ions.] Typically, the susceptible material is a *hardened* steel or stainless alloy. This type of cracking can sometimes occur under aerobic conditions (e.g., as of very hard or highly stressed steel in salt air), but is more usually associated with anaerobic conditions.

Although there are a number of chemical species (e.g., hydrogen cyanide) which poison the dimerization of atomic hydrogen, thereby promoting HAC and other hydrogen-related phenomena, the most commonly encountered agent is hydrogen sulfide. Indeed, the phenomenon is largely known in engineering circles as *sulfide stress cracking* (SSC).

The primary engineering reference for materials selection to resist SSC is the NACE Standard MR-01-75, which defines service parameters, suitable materials, and metallurgical restrictions (e.g., maximum permissible hardness) pertinent to "sour" service.[5] Hydrogen sulfide and derivative chemical species are widely encountered in certain oil and gas operations particularly.

The reduction of stress-level effective in combatting SSC is usually effected by tempering or drawing the hardened metal (e.g., to a Rockwell C hardness less than 22 for alloy steels), rather than by "stress relief" itself.

Because of the mobility and activity of atomic hydrogen, the tendency toward SSC is greatly reduced above about 80°C (175°F).

Liquid-Metal Cracking

LMC is sometimes also known as *liquid-metal embrittlement* (LME), although this is misleading because some metals do not specifically crack during LME.

LMC seems to be a chemical-mechanical effect in which a liquid metal quickly penetrates the grain boundaries under the simultaneous influence of tensile stress. Again, specificity is observed; certain liquid metals attack some alloy systems but not others.

Most often, the liquid metal is in or of the environment, but can also be produced electrochemically (e.g., by cathodic reduction of mercury ions).

Such mercury cracking is typical of high-strength copper-base alloys in mercury salt solutions or by mercury vapor contamination from catalysts, broken thermometers, or blow manometers.

In modern practice, LMC problems more often arise from *molten* metal contamination. Zinc from galvanized hardware or zinc-pigmented paints can cause LMC of 18-8 type stainless steels during welding or heat treatment.

Note: Contrary to the Materials Technology Institute (MTI) manuals 1 and 15, molten *aluminum* does *not* cause LMC of stainless steel;

private and published sources had erroneously reported that it does at the time of writing.

Molten lead and cadmium are LMC agents for austenitic stainless steels, but are less commonly encountered. Cadmium will also cause LMC of steel, as will molten copper (e.g., in welding of copper-clad steel). *Sulfur* contamination of nickel-rich and nickel-base alloys can cause a form of LMC, especially during welding of contaminated surfaces, by penetration of grain boundaries by a nickel-sulfide eutectic.

Conclusion

The three forms of environmental cracking pose specific dangers to process equipment in a variety of services unless their potential is recognized.

These phenomena are best combated by proper materials selection when resistant materials are available and cost-effective. Always, proper design and heat treatment or other metallurgical controls should be implemented to minimize the possibility of cracking failures.

Any type of corrosion failure involving cracking phenomena should be evaluated by a competent corrosion/materials engineer to properly determine the mechanism responsible.

References

1. D. R. McIntyre and C. P. Dillon, "Guidelines for Preventing Stress-Corrosion Cracking in the Chemical Process Industries," manual no. 15, Materials Technology Institute of the Chemical Process Industries, St. Louis, 1985.
2. T. P. Oettinger and M. G. Fontana, "Austenitic Stainless Steels and Titanium for Wet Air Oxidation of Sewage Sludge," *Materials Performance,* vol. 15, no. 11, 1976.
3. F. L. LaQue and H. R. Copson, "Corrosion Resistance of Metals and Alloys," 2d ed., Reinhold, New York, 1963, p. 410.
4. C. P. Dillon, *Corrosion Control in the Chemical Process Industries,* McGraw-Hill, New York, 1986.
5. "Sulfide Stress Cracking-Resistant Metallic Materials for Oilfield Equipment," NACE Standard MR-01-75, NACE, Houston (latest edition).

61

Corrosion
under Insulation

Failures attributable to corrosion phenomena under insulation have been recognized for more than 40 years. Their growing practical importance is such that a full-scale symposium on the subject was held in 1982.[1]

The two major problems are severe general corrosion of steel vessels under wet insulation and external stress-corrosion cracking (ESCC) of the 18-8 stainless steels. A current report covers the use of protective coatings as a preventive measure.[2]

The corrosion phenomena are always associated with *wet* insulation, caused either by failure of weather barriers against storm-driven rain or spray or by leakage of industrial waters above an insulated system, permeating the insulation.

Steel

The external corrosion of steel under wet insulation is a critical problem, both because of its severity and because it is not visible to the naked eye.

Older types of insulation (hair, felt, magnesia, and asbestos) were prone to absorb and hold water by capillary action. Foamed glass and other modern types of insulation are of the closed-cell type and less prone to do so. Nevertheless, water *under* insulation is a powerful corrodent.

Operating conditions contribute to the occurrence and severity of the problem. The worst situations are those in which a distillation column operates at below the freezing point at the top and above ambient temperature in the bottom sections (e.g., in gas separation operations). The top of the column may be at $-10°C$ ($-18°F$), with the base temperature 60°C (140°F); hot water, from melting ice formed by entering moisture, and air constitute the corrosive agents.

Wholly cryogenic operations are at risk during shutdown and where piping or structural members exit the cold-temperature insulated areas. Equipment operating at constant, uniform temperature is at minimum risk. Surface temperatures above about 90°C (195°F) are relatively harmless if the ingress water is mineral-free (i.e., condensation, demineralized water, or steam condensate). On the other hand, mineral-laden waters will simply evaporate and concentrate their salt content, with an attendant increase in corrosivity.

The severity of corrosion of steel under wet insulation can be as high as 8 mm/a (300 mpy). The danger of a pressure release of flammable or toxic gases or other materials, without prior warning, is very real.

Although it is possible to inspect a vessel by providing "windows" in the insulation, this procedure may provide additional weak points for ingress of moisture. It is preferable to coat the vessel with a heavy-duty organic paint system appropriate to the service temperature [e.g., 0.3 to 0.4 mm (12 to 15 mils) of catalyzed epoxy] and maintain the weather barrier.

Note: Do not use zinc-pigmented coatings in this application. The combination of soft-water chemistry and elevated temperature can reverse the potential between zinc and steel, accelerating rather than alleviating external corrosion of the steel substrate.

Stainless Steel

The earliest reported instances of ESCC of 18-8 stainless steels under insulation were ascribed to chloride in the insulation.[2] Some installations had been insulated with magnesia (MgO), which exacerbated the problem, but even silicate-based insulation gave problems in some cases. For some time, this problem was approached by adding corrosion inhibitors to the insulation itself (e.g., sodium silicates) and by specifying a maximum allowable chloride content.

Some cases were caused by the use of PVC coatings intended to provide a scratchproof surface to the insulation. These coatings would hydrolyze in the presence of warm water, providing an acid chloride medium which promoted ESCC. Ashbaugh[3] determined that some of his plant failures arose from storm-driven chloride-bearing water along the Texas Gulf Coast.[3,4] Similar problems arose in Puerto Rico and Louisiana, but incidents have been observed as far north as Nova Scotia. In nonmarine situations, the moisture and chlorides entered the insulation from water leaks (or even deliberate water sprays created to augment cooling or wash down chemical leaks).

Eventually, it was recognized that the most effective way to prevent ESCC was to coat the stainless vessels in a manner analogous to the coating of carbon steel. There are further requirements, however, in that the coatings for stainless steel must be free of hydrolyzable organic chlorides (e.g., PVC; see Chap. 60) and zinc (which poses a potential problem of LMC during welding or fire). An ideal coating system is a modified silicone-based paint appropriate for elevated temperatures, although other types may be used for temperature ranges not requiring the silicone system.

Because of the uncertainties in, and possible changes of, paint formulations, it is recommended that the systems be selected by (and periodically checked with) the wicking test prescribed for this purpose.[4]

There is little surface preparation required (unlike the coating of steel), a solvent wipe being usually sufficient. A two-coat application is suggested for optimum protection.

Economics suggests that vessels and piping of 6-in (152.4 mm) size and above should be so protected. For smaller-diameter piping, it is sufficient to coat for several inches in the flange area, and steel flanges should themselves be coated with a heavy-duty system (e.g., catalyzed epoxy) to prevent rust deposits on the stainless steel surfaces.

References

1. W. I. Pollock and J. M. Barnhart, *Corrosion of Metals Under Thermal Insulation*, STP 880, ASTM, Philadelphia, 1986.
2. "A State-of-the-Art Report of Protective Coatings for Carbon Steel and Austenitic Stainless Steel Surfaces Under Thermal Insulation and Cementitious Fire-Proofing," *Materials Performance*, vol. 29, no. 5, May 1990; NACE publication GH189, item no. 54268.
3. W. G. Ashbaugh, "ESCC of Stainless Steel Under Thermal Insulation," *Materials Protection*, vol. 4, no. 5, May 1965.
4. W. G. Ashbaugh, "Stress Corrosion Cracking of Process Equipment," *Chemical Engineering Progress*, October 1970.

Total Acidity/
Alkalinity vs. pH

There is often considerable confusion relative to the use of pH as a measure of corrosivity. It is generally recognized that acid pH's (e.g., less than 4.5) are corrosive to iron and steel, while pH's greater than 9.5 are much less so. Amphoteric metals (e.g., zinc, tin, and aluminum) may be corroded *either* by acids or alkalis. Acid pH's are not *in themselves* corrosive to metals above hydrogen in the electromotive force (EMF) series (e.g., copper, nickel), because they cannot liberate hydrogen from the acid environment, whereas the more anodic metals and alloys (Mg, Zn, Fe) are attacked.

The pH is the negative logarithm (base 10) of the hydrogen ion concentration. In *strong* acids and bases, it is a valid measure of the acidity or alkalinity, because of the total ionization of the electrolyte. In such cases, a useful mnemonic device is that a pH of 4 is *approximately* 4 ppm acid (specifically 3.65 ppm HCl, 4.9 ppm H_2SO_4, etc.). The obverse is true as well; a pH of 10 is 4 ppm NaOH. Remember also that each pH unit is 10 times stronger than the neighboring value (i.e., pH 3 is 10 times more acidic than pH 4; 36.5 ppm vs. 3.65 ppm HCl). A pH of 12 is 400 ppm NaOH, if that is the contributing species. The *neutral* pH of 7, at which there are equal concentrations of hydrogen and hydroxyl ions, derives from the fact that the *autoprotolysis constant* of water is *14* (see further below).

For weaker acids and alkalies, which are less strongly ionized, the total acidity (or alkalinity) is a better indication of corrosivity. The formation of carbonic acid by absorption of CO_2 in aqueous solutions results in a weakly acid solution, which is much more corrosive to iron and steel than is indicated by a pH of 5 to 5.5, for example. The same is true of organic acids like acetic or propionic, although formic and oxalic acids are quite strong. The weak acids, only slightly ionized, are a *reservoir,* as it were, of potential protons available for corrosion.

Fully ionizable acid salts [e.g., sodium bisulfate ($NaHSO_4$), ferric chloride, ammonium phosphate] function like dilute solutions of their corresponding acids. The pH of a ferric chloride solution, apart from the specific ionic effects of Fe^{+++} and Cl^-, is not a reliable indication of corrosivity, although lower pH's are obviously worse than higher values.

The best way to evaluate acid solutions is to measure *both* total acidity, by titration to pH 7, and pH and chloride ion concentration, for example. One can then determine what components are present and interpret their potential activity. Various anomalies may be observed. For example, soluble iron salts will give an erroneously high total acidity, if titrated to a phenolphthalein end point of pH 8.3, as the ferric hydroxide is precipitated. Formate esters will also titrate as total acidity, although substantially noncorrosive. In both cases, the pH is a better measure of potential corrosivity. On the other hand, total acidity is more indicative of corrosion in sulfurous acid/chloride liquors (e.g., in flue gas desulfurization systems) because sulfurous is a relatively weak acid.

It should be noted also that pH's measured in organic media require much interpretation. In alcohols and glycols, the autoprotolysis constants are 16+, so a *neutral* pH is about 8 to 8.3. In dilute aqueous solutions, this may be unimportant, but not in primarily organic streams. In some organic solvents, pH is essentially meaningless.

In one case involving corrosion of steel evaporators by crude ethylene glycol–water mixtures, not only was the relationship between total acidity and pH obscured by iron precipitation during titration but free carbon dioxide appeared to be the corrosive species. However, installation of a carbon dioxide stripping still failed to alleviate the corrosion. Subsequently, it was determined that there were 6 to 15 ppm of organic acids (e.g., formic, oxalic, glycolic) remaining after CO_2 removal which continued to corrode the steel multiple-effect evaporators because the corrosive species were constantly replenished. Since the process would tolerate neither caustic neutralization nor use of a corrosion inhibitor, replacement in a stainless steel evaporator was required.

The extent to which an organic medium actually dilutes an acid or alkali is important. Small amounts of sulfuric acid in an organic carrier may behave more like *concentrated* sulfuric than dilute, if the solvent has little dilution effect. In complex organic ethers, traces of alkali have been observed to act like hot, concentrated sodium hydroxide, attacking carbon steel directly (although type 304 stainless steel was unaffected).

The conclusion is that corrosion data are a more reliable indication than either pH or total acidity (or alkalinity) in complex process mixtures. However, pH and other relevant chemical features should be ascertained, where possible, as an aid in anticipating potential problems. A professional corrosion/materials engineer with a chemistry or chemical engineering background has a distinct advantage over the corrosion metallurgist in such types of problems.

Microbiological Corrosion

Microbiological corrosion, sometimes abbreviated as MIC or BIC, is *not* a form of corrosion. Rather, some of the eight forms of corrosion[1] are exacerbated by BIC. Biomasses of one kind or another can induce localized corrosion (e.g., concentration-cell attack, as in dissolved oxygen or dissolved sulfur cells or metal-ion cells) by adhering to the metal surface. Some other forms of localized corrosion (e.g., pitting, dealloying, intergranular attack, environmental cracking) are caused either by specific chemical agents produced by metabolic processes in or from the decay of biomasses (e.g., ammonia, sulfides, ferric or manganic chloride), or by electrochemical or physical concentration of chloride ions within a biomass.

BIC may be induced by the direct or indirect activities of bacteria, fungi, algae, and yeasts. Microbes can selectively consume molecular hydrogen (i.e., promote cathodic polarization), oxidize metal ions from a lower to a higher state, or produce ammonia, methane, sulfur, or hydrogen sulfide.[2]

The most common problems arise from bacterial action in soil or water and may be caused either by aerobic or anaerobic bacteria, of which the latter are more common.

Aerobic Bacteria

Any kind of organic matte can induce chloride concentration and/or concentration-cell corrosion. However, there are specific strains of aerobic bacteria which pose particular problems.

Filamentous iron bacteria of the *sphaerotilus natans* strain were discovered in 1833 and occur primarily in sewer systems.

The so-called iron-eating bacteria, *Gallionella* and *Crenothrix,* metabolize dissolved iron or manganese salts, excreting voluminous hydrated oxides, often associated with the chloride salts of the oxidized cations (i.e., ferric and/or manganic chlorides). The dissolved metal salts are often present in mine runoff waters and river water downstream of steel mill operations, as well as well waters. Steel and cast iron are subject to pitting by oxygen concentration-cell attack, while stainless steels are prone to pitting and other oxidizing chloride effects (e.g., intergranular attack, stress-corrosion cracking).

The waste products from these strains are typically a bulky reddish-brown material, having a high loss by ignition and a high iron content (both typically in excess of 35%).

Anaerobic Bacteria

These are most commonly encountered as sulfate-reducing bacteria (SRB), present in most soils and waters. They are indigenous to waters associated with oil-bearing shales and strata.[3]

Biological reduction of sulfur, like nitrogen fixation and oxygen production, is an essential part of life processes. Some such reduction is assimilatory (e.g., green plants utilizing ammonium sulfate fertilizer) while others are dissimilatory (e.g., using sulfate ions like oxygen, without assimilation).

SRB were discovered in 1895, their primary manifestation being graphitic corrosion of gray cast iron water mains by anaerobic water drawn from artesian wells. Technical reviews of the subject were published in 1936, 1945, 1959–65, and 1973. SRB are known to reduce sulfites, thiosulfates, and tetrathionates, as well as sulfates themselves.

Because they are anaerobes, growth of SRB is inhibited by oxygen or air ingress. However, the bacteria are only rendered dormant, *not* killed, and mixed populations may show aberrant behavior. (Also, colonies of SRB may develop *under* deposits, which promote locally anaerobic conditions in an otherwise aerobic environment.) Growth is slow anyway, compared with common soil and water bacteria (e.g., *Pseudomonas*). However, the incidence of SRB is widespread and they readily survive in several subspecies. *Desulfovibrio rubentschickii* reduce acetate ions, while *Desulfuromonas acetoxidans* reduce elemental sulfur (but *not* sulfates or sulfites) to sulfides.

There are both mesophilic and thermophilic varieties of SRB. Among the former, *Desulfovibrio* and *Desulfomonas* will tolerate a temperature from slightly below 0°C (32°F) (freezing kills most of the population) to about 45°C (113°F). The thermophilic *Desulfomaculum* is resistant to about 60°C (140°F) but strains have been reported to service temperatures as high as 90° to 104°C (195° to 220°F). Note that bacterial action has been detected in some geothermal springs.

For SRB to grow as colonies, there must be some organic materials present. True autotrophs will grow in a pure mineral environment, but SRB must assimilate acetate/carbon dioxide (mixotrophy) in the process of reducing sulfates. They also seem to require dissolved iron (ammonia and amines promote growth by solubilizing otherwise insoluble ferrous sulfide), as well as a redox potential in the range of about -100 to -250 mV. Mildly alkaline pH's are also beneficial but acid pH's below about 5.3 (the equilibrium pH for carbon dioxide saturation under air at atmospheric pressure) tend to kill SRB.

In freshwater bodies, *sulfureta* (i.e., zones of sulfate reduction) form the anoxic environment necessary for action of SRB. Decaying debris in lake sediment provides organic matter for sulfate reduction and sulfide formation. At the lowest level to which light penetrates, a layer of anaerobic photosynthetic sulfide-oxidizing bacteria develops. Above this, the water is aerobic, the sulfide concentration is low, and aerobic sulfide oxidation takes place. However, in some locations on some occasions, this upper layer is anaerobic and "poisonous dawn fogs" may occur. Marine eruptions are also known to release hydrogen sulfide from similar sulfureta.

Conventional SRB have a limited tolerance for salinity. However, halotolerant (salt-resistant) and marine strains are very common in estuarial water and seawater, proliferating in polluted or otherwise anaerobic waters. "Red tides" are a result of sulfureta of the *Chromatium* and *Thiopedra* strains.

In the process of reducing sulfate to sulfide, intermediate compounds (e.g., sulfites and thiosulfates or thiosulfites) are formed in a stepwise manner. It happens that coexistence of hydrogen sulfide and sulfur dioxide (or sulfites) results in formation of elemental sulfur and water by their interaction. This can cause concentration cells based on dissolved sulfur. It is for this reason that hydrogen sulfide produced by bacterial action is effectively more aggressive than H_2S added as a reactant (e.g., in heavy water extraction). The DS cell can be easily demonstrated in the laboratory with the same permeable membrane apparatus utilized for the oxygen (DO) concentration cell.

Control of BIC

Bacterial and other aquatic forms of life can be controlled by chlorination (or through use of chlorine dioxide or other environmentally acceptable biocides).

Usually, residual chlorine to the extent of 0.1 to 0.2 ppm is sufficient, although levels of 0.2 to 0.5 ppm on a twice weekly basis (and as high as 2 to 5 ppm in some shock treatments) have been employed. It should be noted that the recommended chlorine residual level for swimming pools is 0.3 ppm. '

Nonoxidizing biocides (e.g., hexamethylene biguanide) are sometimes used in swimming pools and in fire-control water systems
(which tend to become anaerobic in the pressurized piping legs).

Of course, the biocide must not itself be reduced by organic contamination, nor unable to reach the colonies of bacteria, if it is to be effective. This requires that mud, silt, and sediment be kept out of water
systems, or periodically removed by pigging, back-flushing, etc.

Specific Materials

Light alloys

Aluminum and its alloys resist hydrogen sulfide quite well, but are
particularly susceptible to concentration-cell and crevice corrosion.
They are likely to be unsatisfactory in natural waters in the presence
of bacterial action. Even aluminum aircraft fuel tanks have suffered
attack in water layers condensed beneath aviation gasoline.

Cast iron and steel

These are susceptible to graphitic corrosion and pitting, respectively,
in soil or water conducive to BIC.

Stainless steels

The straight chromium and chromium-nickel (18-8) stainless steels,
which depend for their corrosion resistance on a passive oxide film,
are particularly susceptible to concentration-cell attack. Both the dep-
rivation of oxygen in the occluded area and the accumulation of chloride ions tends to promote pitting, IGA, and SCC.

The presence of hydrogen sulfide further weakens the passive film,
promoting sulfide stress cracking of martensitic grades, pitting of
ferritic grades, and SCC of 18-8–type alloys at room temperature (as
opposed to the 50° to 60°C threshold temperature usually cited for water exposures).[4]

Any film or deposit, including biomasses, can occlude and concentrate chloride ions to levels greatly exceeding those in the bulk solution. In combination with hydrogen sulfide or dissolved sulfur, this
poses a pitting danger. Also, aerobic bacteria can form ferric or
manganic chlorides which aggravate the problems of pitting and SCC.

It is commonly assumed that high-performance stainless steels of
the 4.5% Mo variety would resist SRB. However, it has recently been
found that alloy 904L (N08904) suffered rapid BIC within 2 weeks
time in North African seawater.[5] This was apparently due to inade-

quate chlorination control. An evaluation of the 6% Mo grades (e.g., S31254 and N08367) is being conducted, but it is likely that they too may be susceptible under *extreme* conditions.

Copper alloys

Copper alloys discourage attachment of bioforms, because of the toxicity of copper salts. They are, however, attacked by ammonia and by sulfurous products of bacterial action.

In the atmosphere in the vicinity of waterborne bacterial action, the high-strength alloys may suffer SCC by ammoniacal compounds. Hydrogen sulfide forms a voluminous, nonprotective sulfide on copper. The yellow brasses form adherent protective tarnishes in the same type of atmosphere.

Under immersion conditions, the thin sulfide films promote localized attack. For example, 90-10 cupronickel (C70600) tubes in aerated seawater may perform well for years, then fail rapidly after a brief exposure to quiescent polluted seawater, because of SRB activity, as during annual plant turnaround.

Nickel alloys

Nonchromium nickel-base alloys (e.g., N04400, N02200) may suffer adverse effects by SRB. The chromium-bearing alloys such as N06600 are susceptible to concentration-cell attack. The chromium-molybdenum grades (N10276 and its homologs) are expected to be resistant.

Reactive metals

Titanium is apparently unaffected by bacterial corrosion in natural waters.

Noble metals

Silver, gold, and platinum are unaffected by bacterial corrosion, except for the tarnishing effect of sulfides on silver.

Nonmetals

Such nonmetallic materials as elastomers, plastics, carbon, and ceramics are unaffected by bacterial action.

Concrete and cement, however, are liable to attack by sulfate ions produced by oxidation of lower sulfur compounds, as well as by any acidic species. The vapor space of sewer pipes is notorious for corrosion of cement-lined or cementitious pipe, because of SRB activity in the wastewater and air-oxidation of hydrogen sulfide in the vapor phase.

References

1. C. P. Dillon (ed.), *The Forms of Corrosion—Recognition and Prevention,* Corrosion Manual No. 1, NACE, Houston, 1982.
2. G. Kobrin, "Corrosion by Microbiological Organisms in Natural Waters," paper no. 147, *Corrosion '76,* NACE, Houston.
3. J. R. Postgate, *The Sulfate-Reducing Bacteria,* Cambridge University Press, Cambridge, U.K., 1979.
4. C. P. Dillon and D. R. McIntyre, "Guidelines for Preventing Stress-Corrosion Cracking in the Chemical Process Industries," publication no. 15, Materials Technology Institute of the Chemical Process Industries, St. Louis, 1985.
5. P. J. B. Scott and M. Davies, "Microbiologically Influenced Corrosion of Alloy 904L," *Materials Performance,* vol. 28, no. 5, May 1989, p. 57.

Carbamate Corrosion

There is a particular problem associated with contamination of ammonia streams by carbon dioxide, with formation of corrosive ammonium carbamate, which often goes unrecognized in chemical processes.

It has long been recognized that the cause of corrosion in urea synthesis is neither the feedstream components (CO_2 and NH_3), which are innocuous to stainless steels, nor the final product of urea (N_2H_4CO). There is a corrosive intermediate, known as ammonium carbamate:

$$2NH_3 + CO_2 \rightarrow NH_2COONH_4 \rightarrow NH_2CONH_2 + H_2O$$

The carbamate or its dissociated acid is responsible for corrosion of stainless steels and even titanium and zirconium under some conditions of operation (e.g., surface iron contamination, and even perhaps contained iron above about 0.05%, can provide a window for hydrogen penetration and attendant hydriding).

Because urea production takes place at elevated temperatures and pressures, ammonium carbamate is present in large amounts. It is not generally recognized that carbamates can form in significant amounts (from the *corrosion* rather than the production standpoint) under much less stringent conditions. The amount formed, and the extent to which it corrodes steels and stainless steels, is a function of temperature, pressure, and concentration. Some typical cases from actual tests and field experiences are as follows:

- When steel coupons were hung in the vapor space of 28% ammonium hydroxide, normally a low-corrosion condition, rates of about 20 mpy were caused by passing a stream of CO_2 through the vapor space. This indicates formation and replenishment of ammonium carbamate.

- Corrosion of both steel column walls and type 410 (S41000) valve trays were observed in an ammonia stripping still in an alkylamines process (see Chap. 39). The corrosion was occasioned by carbamates formed when CO_2 was accidentally formed in the catalyst bed. There was *no* corrosion above the feed in dry ammonia, nor below the corroded area in wet amines/water solutions. Attack was confined to the area of transitory ammonium (and possibly amine) carbamate formation.

- Corrosion of a type 304L (S30403) ammonium carbonate column was observed intermittently in a fertilizer plant. This attack was duplicated in a laboratory column kettle, in both liquid and vapor exposures, by charging a completely deaerated feed to the system in a 16-hour test. When the feed was *not* deaerated, the stainless steel was unattacked, showing the passivating effect of dissolved oxygen. (*Note:* Oxygen has long been blown into urea reactors to minimize corrosion of type 316L and even higher stainless alloys and reactive metals like titanium and zirconium.)

- A type 316L (S31603) ammonia column in a methyl ethyl pyridine process suffered rapid corrosion because no provision had been made to purge the column. A gradual increase in carbamate concentration arose from ingress of trace amounts of CO_2 in the reaction system.

- A steel/low-alloy steel compressor handling carbon monoxide odorized with methylamine suffered unexpected corrosion. This was apparently due to trace amounts of carbon dioxide in the CO, and attendant formation of carbamate and traces of water from dissociation of the carbamate.

These tests and observations indicate that at least small amounts of carbamates may be formed whenever ammonia or amines are contaminated by trace amounts of carbon dioxide. Unexpected corrosion of ammonia and amine systems had been previously observed, but unexplained, in a variety of processes at several chemical plants, primarily with steel systems having 13% chromium-grade stainless steel accessories intended to cope only with ammonia/amine corrosion.

The author also investigated a problem involving corrosion of a steel ammonia recovery system in which *no* carbon dioxide contamination was possible. The miscreant appeared to be traces of carbon disulfide (CS_2) solvent carried over from the process. This would form *thiocarbamates* (NH_2CSSNH_4) in a manner directly analogous to the ammonium carbamate reaction.

Similar compounds may be inherent in the corrosion processes involved in acid-gas removal systems wherein carbon dioxide and/or hydrogen sulfide are preferentially absorbed in alkanolamine solutions. If so, however, they would be inherent in the absorption reaction and would necessarily be combated by corrosion inhibition and/or materials selection, as is standard practice.

Sensitization and "Weld Decay"

The problems associated with sensitization of austenitic stainless steels and related alloys, including localized corrosion of the heat-affected zone of welds, seem to be widely appreciated but as widely misunderstood in many instances.

The nature and mechanisms of sensitization of conventional 18-8 stainless steels have been studied and reported since the 1930s. Modern technology has given us compositional and other quality-control techniques which should obviate the problem, yet errors of both commission and omission continue. The following discussion is intended to clarify some of the conditions under which sensitization and its related phenomena are (or are not) a problem in the CPI. A brief review of the problem is a necessary prelude to the discussion.

Sensitization Mechanisms

Broadly speaking, sensitization is a phenomenon in which, by holding a metal or alloy in a specific temperature range for some prolonged period of time, metallurgical changes occur which render it susceptible to *intergranular* corrosion in specific environments. Stainless steels, nickel-rich and nickel-base alloys, and zirconium are notoriously susceptible to this phenomenon under specific conditions of heat treatment or fabrication and chemical exposure.

The phenomenon occurs most commonly in the austenitic stainless steels. When a regular-carbon grade, such as type 304 (S30400), is held in the temperature range of 425° to 815°C (800° to 1500°F), carbon migrates to the grain boundaries and precipitates as chromium carbides (probably Cr_6C_{23}). This leaves each grain surrounded by a chromium-depleted zone of inferior resistance compared with the matrix. Intergranular corrosion occurs in environments such as oxidizing acids (e.g., nitric acid or sulfuric acid containing ferric or cupric ions), while intergranular stress-corrosion cracking (IGSCC) occurs in supercritical water, as in nuclear-powered boilers.

When the phenomenon is observed in the heat-affected zone (HAZ) of welds (which are between the hot molten weldment and the cold parent metal), the phenomenon has been called *weld decay*. Note that this is a misnomer, in that it is usually *not* the weld itself that suffers IGA. Of course, sensitization can also occur during improper annealing (too slow a quench) or during hot-working or stress-relieving operations; this will affect the entire metal structure.

The problem as originally encountered was combated by addition of elements which combine preferentially with carbon, in lieu of the chromium. Columbium and titanium additions at 5 to 10 times the carbon content have been used to change type 304 to types 347 (S34700) and 321 (S32100), respectively.

Technological improvements since the early 1950s, as well as more recent AOD processes, permit the manufacture of alloys of very low carbon content, such as S30403 and S31603. When carbon contents are held to such low levels, chromium carbide precipitation still occurs during sensitization but in such small amounts during short-time heating, in welding, for example, as not to affect corrosion resistance. It should be observed that the low-carbon grades may *not* be substituted for the stabilized grades when service conditions include long-time exposure to sensitizing temperatures.

Low-carbon stainless steels of both the 18-8 variety and high-performance grades are standard construction materials for the CPI, normally resistant to sensitization during ordinary welding and fabrication.

Practical Problems

Improper specification

It is not necessary to specify low-carbon grades *unless* the environment is specific for IGA. Obviously, any error is on the conservative side and it may be well to use them if there is any possibility that conditions conducive to IGA *might* arise. However, low-carbon grades are *not* required for storage and handling of refined chemicals and high-quality waters. (High-chloride waters, especially brackish or marine waters, cause accelerated *pitting* in the sensitized zones, even when IGA does not occur.)

As a practical matter, low-carbon stainless piping is the industry standard; regular-carbon piping is often that which just fails to meet low-carbon chemistry. Sheet and plate for vessels may be specified either way, as required. For innocuous service, piping should be specified as "S30400 (S30403 acceptable)."

Quality control

In those instances in which a low-carbon or stabilized grade is required, it is unwise to rely on the mill analyses as an indication of resistance. The analysis may be in error or local or surface contamination may have occurred (see "Improper handling and fabrication" below). Proper specification should include an appropriate quality assurance corrosion test (e.g., ASTM A-262) to *verify* that the material will indeed resist IGA.

Improper handling and fabrication

A properly constituted low-carbon grade may become carbon-contaminated during heat treatment or fabrication. Mill defects should be picked up by qualification tests. Carbon pickup during welding or fabrication (e.g., from oil or grease, from improper gas shielding) can be eliminated only by proper cleaning and care in handling and welding.

Although wrought products in the low-carbon and stabilized grades do not require a full anneal, *castings* (e.g., CF3M) should be solution-annealed to ensure homogeneity.

Titanium-stabilized grades (S32100) require a "stabilizing anneal" at about 950°C (1750°F) to ensure reaction of the titanium with the carbon. Columbium-stabilized grades do not require this heat treatment, the columbium scavenging carbon during the melting process.

Knife-line attack

Unfortunately, titanium and/or columbium carbides can be redissolved by the heat of welding, undoing the stabilizing effect. Cross-welds and multipass welds may heat a small area in the fusion zone above about 1260°C (2300°F), redissolving columbium carbides and permitting sensitization in that area. The attendant IGA of the fusion zone is called *knife-line attack* (KLA).

The higher the nickel content, the lower the temperature for dissolution of columbium and titanium carbides. Alloys like N08020 and N08825 are subject to KLA when heated to about 980°C (1800°F). They should not be hot-worked in that temperature range. Low-carbon high-performance alloys (e.g., N08904, N08367, S31254) are preferred where otherwise suitable.

Other phases

There are constituents other than chromium carbides which can cause IGA. Ferrite segregation can be selectively attacked by reducing acids. The thermal conversion product of ferrite, sigma, is selectively attacked by oxidizing acids. This type of attack can be studied in appropriate quality-assurance tests.

In molybdenum-bearing grades (S31603), chi phase can cause IGA. In a pre-precipitation state, it is responsible for IGA of *annealed* (let alone sensitized) material in boiling 65% nitric acid. The nitric-HF test or various sulfuric acid/oxidizing ion tests per A-262 et al. must be used to evaluate type 316L and related alloys.

Service failures have occurred by IGA in fully qualified 316L-clad steel vessels because of prolonged stress-relief of the vessels. The thermal stress-relief cycle took several days for these heavy-walled vessels [heating to 650°C (1200°F) and slow cooling], and they subsequently failed by IGA over the entire surface in hot acetic acid service, because of chi-phase corrosion.

Other alloys

Sensitization and IGA have been encountered in the original nickel-molybdenum and nickel-chromium-molybdenum alloys, but modern variants (N10665, N10276, N06022) are substantially immune to these phenomena during normal fabrication.

Zirconium is subject to IGA in HCl contaminated with oxidizing ions, notably Fe^{+++}. This can probably be prevented only by avoiding specific services of this type.

66

Pyrophoric
Corrosion Products

Pyrophores (from the Greek for *fire makers*) are materials which will react exothermically with air or oxygen. It is well-known that finely divided metals tend to do this, even when more massive pieces will not burn or support combustion, because of a large area-to-mass ratio in powder form. Using a grinding wheel on titanium or zirconium yields a brilliant pyrotechnical display as particles of reactive metal burn.

Powdered magnesium has been used in flares and fireworks, and aluminum powder is used in thermite reactions with iron oxide powder (e.g., in brazing copper wires to steel pipe for cathodic protection connections). Aluminum prilling towers for molten sulfur have ignited and burned, apparently as the result of abrasion by rusted steel structural members. Aluminum-pigmented paint on cast-iron steam radiators has become susceptible to sparking by scratching. Turnings of titanium and zirconium have caused fires in machine shops because of spontaneous ignition.

Not so well-known is the possibility of pyrophoric corrosion products, although they constitute a definite ignition hazard and thus create potential for fire or explosion in process operations. Iron or steel can form pyrophoric sulfides, and zirconium and titanium form pyrophoric compounds under certain conditions. A few other ignition and explosion hazards are known (e.g., copper acetylides, silver azides from decomposition of ammoniacal silver nitrates, organic peroxides), but they are inherently unstable, rather than reactive with air, are not usually corrosion-related, and are usually of lesser practical import.

Iron and Steel

Pyrophoric ferrous sulfide is formed under some conditions involving exposure of iron or steel to wet hydrogen sulfide and is a problem in some sour oil and/or gas operations. On exposure to air, as in a newly opened cargo tank, pressure vessel, or pipeline, the ferrous sulfide glows red-hot and the iron salt is oxidized to ferric oxide with liberation of sulfur dioxide fumes. Any flammable liquids or air-vapor mixtures can be ignited by the red-hot compounds. In some cases, red-hot surfaces have developed when cast-iron pipes have been excavated. Although this has been ascribed in some cases to the fine graphite particles left as a residue from graphitic corrosion, it is most likely attributable to, or exacerbated by, sulfidic residues from bacterial (i.e., SRB) corrosion in the soil (see Chap. 63).

It has been postulated that formation of pyrophoric iron sulfide requires certain preexisting oxide products (e.g., suboxides of iron, such as Fe_3O_4 or other ferrous oxides), but this has not been definitely established.

Damage from such sulfides has varied from burned tires (or even whole vehicles) when trucks were parked on materials pigged from oil pipelines, to fires and explosions in refineries or tanker-ship cargo compartments.

It is impossible to prevent the formation of pyrophoric iron sulfides when iron or steel are exposed in sour service. The only practical protective measure consists of gas-freeing or inerting vessels and piping before exposure to air. Iron sulfides of any kind in small vessels can be easily destroyed by cleaning with inhibited hydrochloric acid but this may be impractical in extensive piping or equipment.

Zirconium

An industrial accident circa 1985 burned a maintenance worker when a zirconium heating coil from a sulfuric acid pickling bath heated up on exposure to air. A safety alert was issued suggesting that *all* zirconium vessels be labeled potential fire hazards! Fortunately, the Materials Technology Institute of the Chemical Process Industries noted that the phenomenon involves only finely divided metal or corrosion products, such as those from intergranular corrosion, and issued a technical report on the subject.[1] There had been numerous similar incidents in the nuclear industry about 30 to 35 years previously.[2,3]

A survey of the chemical and petrochemical industry indicates that pyrophoric zirconium corrosion products arise mostly from corrosion by sulfuric or hydrochloric acid streams contaminated with oxidizing ions (Fe^{+++}, Cu^{++}). The incidence is low, but the potential hazards are real enough. The MTI documents describe maintenance cleaning measures designed to minimize the hazard.

Titanium

The pyrophoric tendencies of titanium are well-known in the nuclear field, where ignition of titanium pump impellers and of titanium trim in valves has been reported.[2,3]

In the chemical industry, titanium has caught fire in wet chlorine when insufficient water content was present to sustain passivity and in a fuming nitric acid pump run against a dead leg, which permitted the temperature of the acid to rise to the point where ignition of the metal occurred. Titanium distillation equipment in sour service has reportedly been ignited by pyrophoric iron sulfides. One petroleum company reported that prior seawater/sour service predisposed titanium heat-exchanger tubes to ignition by a torch application in the shop.

The engineer concerned with materials of construction for chemical service should be aware of the potential for fire or explosion due to the possible presence of pyrophoric corrosion products.

References

1. C. P. Dillon and D. R. McIntyre, "Pyrophoric Behavior and Combustion of the Reactive Metals," publication no. 32, Materials Technology Institute of the Chemical Process Industries, St. Louis, 1988.
2. J. C. Griess et al., "Report on Corrosion of Titanium Impeller No. 1," report ORNL-CF-55-9-55, Oak Ridge National Laboratory, Oak Ridge, Tenn., 15 September 1955.
3. J. P. Hammon et al., "Failures of Titanium Alloy Trim in HRP Dump Valve Loop," report ORNL-CF-56-8-214, Oak Ridge National Laboratory, Oak Ridge, Tenn., 15 August 1956.

Index

Note: Because of the nature and content of this book, no attempt has been made to index materials in detail. Corrosion behavior of various relevant materials is found under appropriate processes or environments. The index covers processes, specific chemicals, and specific phenomena.

ABOUT THE AUTHOR

C. P. Dillon is a certified NACE corrosion specialist and a registered professional engineer with more than 42 years of experience in corrosion engineering. He is a recognized authority on corrosion control in industry and specializes in the resolution of the special problems associated with chemical processes, oil and gas treatment, and the treatment of industrial waters. As president of C. P. Dillon & Associates in St. Albans, West Virginia, Mr. Dillon is currently a respected and sought-after consultant and lecturer.